Erwin Schrödinger

Our Understanding of Matter and the Universe

Edited by Vesselin Petkov

MINKOWSKI
Institute Press

Erwin Schrödinger
12 August 1887 – 4 January 1961

Cover: `https://www.spitzer.caltech.edu/system/avm_images/binaries/6681/huge/ssc2019-15d.jpg?1603782341`

ISBN: 978-1-989970-31-7 (softcover)
ISBN: 978-1-989970-32-4 (ebook)

Minkowski Institute Press
Montreal, Quebec, Canada
http://minkowskiinstitute.org/mip/

For information on all Minkowski Institute Press
publications visit our website at
http://minkowskiinstitute.org/mip/books/

EDITOR'S PREFACE

This volume contains new publications of two essays by Schrödinger "Our Image of Matter"* (1952) and "What is an Elementary Particle?"† (1957) and his book *Expanding Universes*‡ (1956).

The idea of this collection of works by Schrödinger was to present his views on the whole world – from the quantum level to the Universe. This was achieved at a price – unlike the first two essays the third work – the book *Expanding Universes* – is technical.

The texts were typeset in LaTeX and noticed typos were corrected. A number of editor's notes were made in the third work.

27 November 2021

Vesselin Petkov
Minkowski Institute
Montreal

---

*Lecture presented on September 4, 1952. It was originally published in Italian as Chapter 3 in the collection Heisenberg, Born, Schrödinger, Auger, *Discussione sulla fisica moderna* (Paolo Boringhieri Editore, Turin 1960). The English translation of the book by M. Goodman and J. W. Binns appeared two years later – Heisenberg, Born, Schrödinger and Auger, *On Modern Physics* (Collier Books, New York 1962).

†Published as Chapter IX in E. Schrödinger, *Science Theory and Man* (George Allen and Unwin Ltd 1957).

‡Original publication: Erwin Schroödinger, *Expanding Universes* (Cambridge University Press, Cambridge 1956)

CONTENTS

Editor's Preface   i

Our Image of Matter   1
1. The crisis – a preview 1
2. Some observations on corpuscles 3
3. Wave field and particle: their experimental demonstration 6
4. The quantum theory: Planck, Bohr, de Broglie   9
5. Wave field and particle: their theoretical relationship 11
6. Quantum steps and identity of particles 15
7. Wave identity 17
8. Conclusion 19

What is an Elementary Particle   21
1. A Particle is not an Individual 21
2. Current Views: the Amalgamation of Particles and Waves 22
3. Current Views: the Nature of Waves 23
4. Current Views: the Nature of Particles (Uncertainty Relation) 24
5. Current Views: the Meaning of the Uncertainty Relation 26
6. Criticism of the Uncertainty Relation 27
7. The Notion of a Piece of Matter 30

8. Individuality or 'Sameness' 31

9. The Bearing on Atomism 32

10. The Meaning of the New Statistics 33

11. Restricted Notion of Identity 37

12. Crowdedness and Wave Aspect 38

13. The Condition for the Particle Aspect 39

EXPANDING UNIVERSES   43

Preface   45

I. The de Sitter Universe   47

1. Synthetic construction 49

2. The reduced model. Geodesics 51

3. The elliptic interpretation 56

4. The static frame 64

5. The determination of parallaxes 73

6. The Lemaître-Robertson frame 82

II. The Theory of Geodesics   95

7. On null geodesics 97

   (a) Determination of the parameter for null

         lines in special cases 100

   (b) Frequency shift 105

8. Free particles and light rays in general, ex-

   panding spaces, flat or hyperspherical 111

   (a) Flat spaces 112

   (b) Spherical spaces 115

   (c) The red shift for spherical spaces 122

III. Waves in General Riemannian Space-time 124

9. The nature of our approximation 125

10. The Hamilton-Jacobi theory in a gravita-

   tional field 126

11. Procuring approximate solutions of the Hamilton-
      Jacobi equation from wave theory 130

IV. Waves in an Expanding Universe   136
   12. General considerations 137
   13. Proper vibrations and wave parcels 141

Bibliography   159

v

OUR IMAGE OF MATTER[1]

1. The crisis – a preview

The title of this lecture was suggested to me (in its French version) by the Committee. I accepted it gladly. However, before I attempt to do justice to it as well as I can, there are two things that I must say in advance. In the first place, the physicist can today no longer make a significant distinction between matter and anything else in his field of research. We no longer consider forces and fields of force as different from matter; we know that these concepts must be merged into one. To be sure, we say that an area of space is free of matter; we call it empty, if there is nothing present except a gravitational field. However, this is not found in reality, because even far out in the universe there is starlight, and that *is* matter. Further more, according to Einstein, gravity and mass are analogous and therefore not separable from each other. Our subject today is therefore the collective image that physics has of spatiotemporal reality.

---

[1]Lecture presented on September 4, 1952. It was originally published in Italian as Chapter 3 in the collection Heisenberg, Born, Schrödinger, Auger, *Discussione sulla fisica moderna* (Paolo Boringhieri Editore, Turin 1960). The English translation of the book by M. Goodman and J. W. Binns appeared two years later – Heisenberg, Born, Schrödinger and Auger, *On Modern Physics* (Collier Books, New York 1962).

The second point is: This image of material reality is today more unsettled and uncertain than it has been for a long time. We know a great many interesting details; every week we learn new ones. However, to pick out from the basic concepts those that have been established as fact and to construct from them a clear and easily understood framework of which we could say: this is certainly so, this we all believe today, is impossible. There is a wide spread hypothesis that an objective image of reality in any previously believed interpretation cannot exist. Only the optimists among us (and I consider myself one of them) consider this a philosophical eccentricity, a desperate measure in the face of a great crisis. We hope that the vacillation of concepts and opinions signifies only an in tense process of transformation, which will finally lead to something better than the confused series of formulas that today surround our subject. It is for me – but also for you, my respected listeners – most annoying that the image of matter that I am expected to build before you does not yet exist, that there are only bare fragments of a more or less partial factual value. A consequence of this is that in this kind of narrative we cannot avoid contradicting at a later point what was said earlier. This is somewhat like Cervantes, who allows Sancho Panza to lose his beloved small donkey on which he is riding, but a few chapters later has forgotten this, so that the good animal is again with us. In order to avoid a similar reproach, I want to draw up a plan of campaign. I shall report later how Max Planck discovered, over fifty years ago, that energy can be transmitted only in indivisible amounts of definite size – the quanta. Since, however, Einstein soon afterwards proved the identity of energy and mass, we are obliged to say to ourselves that the smallest particles of matter, the atoms or corpuscles, which we have known for a very long time and whose existence is demonstrated today in many elegant experiments in a perfectly "palpable" manner, are just quanta of energy and, so to speak, predate Planck's discovery by more than two thousand years. Because of this it seems even more secure. Here a side glance will be offered on the enormous significance of this discreteness or *countability* of all that exists and happens; only thus can Boltzmann's famous statistical theory of the *irreversible* course of nature really be feasible and clearly understandable.

All this is well and good and certainly contains a great deal of truth. But then Sancho Panza's donkey will return – after more than two thousand years. For I must ask you to believe neither that corpuscles are individuals stable in time nor that the transfer of a quantum of energy from one carrier to another takes place in distinct steps. Discreteness is no doubt involved, but not in the traditional sense of discrete individual particles and certainly not as a discontinuous event. For this would contradict experience from another quarter. The discreteness originates merely as a construct from the laws that govern events. These are still by no means completely understood; but a probably pertinent analogy from the physics of tangible bodies is the manner in which the individual partial tones of a bell result from the finite form of the bell and the laws of elasticity, with which no discontinuity is really connected.

2. Some observations on corpuscles

Let us now begin. The view, already advocated by Leucippus and Democritus in the fifth century B.C., that matter is built up of very small particles, which they called atoms, assumed at the turn of the last century a very definite form as the *corpuscular theory* of matter, entering into interesting details, which became continually clearer and more established during the first decade. Just to outline briefly the fine and fundamentally important individual discoveries made along the way would require two hours of your attention.

The beginning was made by chemistry. Even today there are those who are haunted by the idea that chemistry is the sole and original domain of "atom" and "molecule." From the very hypothetical, somewhat anemic role they played there – the school of Ostwald rejected them flatly – they were for the first time raised to physical reality in the gas theory of Maxwell and Boltzmann. In a gas, these particles are separated by wide spaces, but they are in vigorous motion, they collide again and again, are repelled from each other, and so forth. An accurate pursuit of these processes in thought led first to a full understanding of *all* properties of the gases, elastic and thermal properties, internal friction, thermal conductivity and diffusion, but at the same time to a firm basis for the mechanical theory of heat as a movement of these very small particles, which continually becomes more vigorous with increasing temperature. If this is true, then small bodies that are just visible under the microscope must also be kept in continuous motion by the impact of surrounding molecules, and this motion must increase with rising temperature. This motion of small suspended particles was discovered by Robert Brown (a London physician) as early as 1827, but only in 1905 did Einstein and Smoluchowsky show that it conforms quantitatively to expectations.

In this fruitful period, approximately ten years before and after the turn of the century, so much that was closely related to our subject occurred that it will be difficult to keep it simultaneously in view. Roentgen rays – "light" of very short wave length – and cathode rays – streams of negatively charged particles, the electrons – were discovered. There was the radioactive disintegration of atoms and the radiations emitted during this process – partly

4

streams of particles, precisely those during whose spontaneous emission from the bonding of the atomic nucleus the transformation of one atom into another takes place, and partly "light" of even shorter wave length, which originates during this process. All particles carry an electric charge; the charge is always the very small electric unit charge, measured directly by Millikan, or nearly exactly double or three times it. The mass of these particles could also be measured very accurately, as could that of the atoms themselves.

The determination of the mass of the atoms, so-called mass spectrography, was carried to such fantastic precision by Aston at Cambridge that he was able to answer in the negative a very old question with absolute certainty: they are *not* whole-number multiples of a very small unit. In spite of this, we must imagine these, or more precisely, the heavy but very small positively charged atomic nuclei – the surrounding negative electrons weigh almost nothing – as built up of a number of hydrogen nuclei (protons), of which of course about half have lost their positive unit charge (neutrons). In a normal carbon atom there are, for example, 6 protons and 6 neutrons. It weighs, in a unit convenient for the comparison

$$\text{Carbon atom} \qquad 12.00053\dots$$
$$\text{compared with}\begin{cases} \text{Proton} & 1.00758\dots \\ \text{Neutron} & 1.00898\dots \end{cases}$$

The unit is $(1.6603 \pm \dots)\, 10^{-24}$ gram; this, however, does not interest us for the moment. How do we explain the "mass defect," which in our example amounts to almost a tenth of a unit? By the heat of binding, which is liberated during the combination of these twelve particles and which is enormously greater in such "nuclear reactions" than in the well-known chemical reactions. In other words, the system loses potential energy, as the twelve particles yield to the energy of attraction by which they are held closely together afterwards. According to Einstein, as already mentioned above, this loss of energy signifies a loss of mass. This is called the packing effect. Moreover, the forces are of course not electric ones – these are indeed repellent – but so-called nuclear forces, which are much stronger but act only at very small distances (about 10^{-13}cm.).

3. Wave field and particle: their experimental demonstration

Here you catch me in a contradiction, because I said at the very beginning that today we no longer assume that forces and fields of force are something different from matter. I could easily excuse myself by saying: the field of force of a particle is calculated with the particle. But it is not so. The proved opinion today is rather that every thing – *really everything* – is simultaneously particle and field. Everything has the continuous structure that is familiar to us from the field, as well as the discrete structure familiar to us from the particle. Expressed so generally, this recognition contains certainly a great deal of truth. For it is based on innumerable experimental data. Opinions vary as to details, and of this we shall speak later.

Moreover in the particular case of the field of nuclear force, the particle structure is already fairly well known. The so-called π-mesons, which appear among others during the destruction of an atomic nucleus and which clearly leave behind separate tracks in a photographic emulsion, very probably correspond to it. The nuclear particles themselves, the nucleons – the name under which protons and neutrons are lumped together – which we have been brought up to consider as discrete particles, for their part produce interference patterns in other experiments, when a great many of them are driven against a crystal surface. These patterns leave no doubt that the nucleons also have a continuous wave structure. The difficulty, which is equal in all cases, of combining these two widely different characteristics in one mental picture is still today the major obstacle that makes our image of matter so variable and uncertain.

However, neither the concept of particles nor that of waves is hypothetical. I mentioned in passing the tracks in the photographic emulsion, each indicating the path of a single particle. Known even longer are the tracks in the so-called cloud chamber of C.T.R. Wilson. From these tracks we can observe exceptionally diverse and interesting details regarding the behavior of the single particles, and we can measure them. The bending of their paths in a magnetic field (because they are electrically charged); the mechanical laws in a collision, which takes place approximately as in the case of ideal billiard balls; the destruction of a larger atomic nucleus by means of a direct hit by one of those "cosmic" particles that come from the universe, to be sure in small numbers, but with a tremendous impetus of the single particle, often several million times greater than otherwise observed or artificially produced. There have been efforts in recent times to achieve this at the cost of enormous expenditure, chiefly financed by defense ministries. To be sure it is not possible to shoot anyone with such glancing particles, otherwise we would all now be dead. This study, however, indirectly promises the stepped-up realization of the plan to exterminate mankind, which is close to all our hearts.

It is perhaps as well to say that these interesting observations on single articles, which I cannot wholly describe in my short résumé, can be made only on very rapidly moving particles. The method of the tracks is moreover not the only one. You can easily try out the oldest method yourself, if, one evening in the dark after having got used to the darkness, you examine a luminous number of your wrist watch with a magnifying glass. You will find that it is not uniformly bright, but that it surges and undulates just as the sea often sparkles in the sun. Each scintillating flash is produced by a so-called alpha particle (helium nucleus) emitted by a radioactive atom, which at the same time is transmuted into another one. And this continues in the same manner for a great many years – in a good Swiss watch. Another much used apparatus for the study of cosmic rays is the Geiger-Müller counter which "answers" when it is struck by a single active particle. This is very valuable. Today it is possible with quite familiar methods to amplify this "answer" in such a way that it can release the mechanism of a cloud chamber and the shutter of a camera trained on it at precisely the moment when there is something interesting to photograph in the chamber. This is an important use for these chambers but not the only one. Fifty or more of them are often built in a single apparatus in a complex circuit.

So much for the observation of single particles. Now for the continuous field or wave characteristic. The wave structures of visible light is rather crude (wave length about two-thousandths of a millimeter). It has been very thoroughly investigated for more than a century by means of the effects that occur when two or more or a great many waves cross: diffraction and interference phenomena. The most elegant means for the analysis and measurement of light waves is the diffraction grating, an immense number of fine parallel lines, cut at small equal intervals on a metal mirror, on which light impinging from *one* direction is scattered and recombined in various directions, depending on its wave length.

For the much, much shorter waves of the roentgen

spectrum, as well as for "waves of matter" – high velocity streams of particles – the finest diffraction gratings we can cut are somewhat coarse. In the year 1912 Max von Laue discovered the instrument that has since made possible the exact analysis of all these waves, discovered it in naturally grown crystals. This discovery was invaluable, unique of its kind. Not only does it reveal the structure of the crystal – a highly regular arrangement of atoms with the same groups repeated a countless number of times, each at equal intervals in three directions, "length," "width" and "height" – but this discovery was *one* with the use of the periodic fine structure of the crystal for the analysis of waves in place of the diffraction grating. And of course remember this: the natural structure of the crystal comes to our aid exactly at the point where it, that is, the crystalline structure of matter, makes all precision techniques impossible. It would be impossible to cut gratings of such fineness, because the "material" is too coarse. With these crystal gratings the wave nature of roentgen rays was first established and their wave length measured, and later that of the waves of matter, especially electron streams, but also streams of other particles such as neutrons and protons.

4. The quantum theory: Planck, Bohr, de Broglie

I have now told you many things about the structure of matter, but we have not yet spoken of Max Planck and his quantum theory. Everything I have reported so far could very well have taken place even without it. How then did it happen? What is this quantum theory all about? I shall not relate the exact historical events, but instead tell you how the matter seems to us today.

Planck tells us in 1900 – and the essentials are still

9

true today – that he can understand radiation from red-hot iron or a white-hot star, for example the sun, only if this radiation is produced in portions and is transferred from one carrier to another (for example, from atom to atom) in portions. This was astonishing, because this radiation involves energy, which was originally a highly abstract concept, a measure of the reciprocal action or effective action of these very small carriers. The partition into definite portions was highly surprising not only to us, but also to Planck. Five years later Einstein told us that energy possesses mass and that mass is energy, that they are one and the same – and this too has remained true to the present day. Thus our eyes were opened: our dear familiar atoms, corpuscles, particles are Planckian energy quanta. *The carriers of these quanta are themselves quanta.* It makes your head turn. We realize that here is something fundamental that is still not understood. As a matter of fact, our eyes were not opened suddenly. It took twenty or thirty years. And perhaps they are not altogether open even today.

The direct consequence was less far-reaching but still important enough. In 1913 by means of an ingenious and obvious generalization of Planck's statement Niels Bohr taught us to understand the *line spectra* of atoms and molecules and simultaneously the construction of these particles from heavy, positively charged nuclei and light electrons circling around them, each of them carrying a unit negative charge. I must here shirk a detailed explanation of this important transition stage in our knowledge. The fundamental idea is that each of these small systems – atom or molecule – can harbor only definite, *discrete* quantities of energy corresponding to its nature or structure; that during a transition from a higher to a lower "level of energy" it emits the excess as a radiation quantum of quite definite wave length, which is *inversely proportional* to the quantum given up (this was already included in Planck's original hypothesis).

This implies then that a quantum of a given amount manifests itself in a periodic process of quite definite *frequency*, which is *directly proportional* to the quantum (the frequency is equal to the energy quantum, divided by Planck's famous constant h). The really quite obvious deduction, that with a particle mass m, which according to Einstein has an energy mc^2 (where c = the velocity of light), there must be associated a wave process of frequency mc^2/h was first drawn by *L.* de Broglie in the year 1925, first for the mass m of the electron. Only a few years after this famous doctoral thesis of de Broglie, the theoretical "electron waves" required in his postulate were experimentally demonstrated, in the manner I have already discussed. This was the point of departure for the early recognition, also already mentioned, that everything – *really everything* – is both particle and wave field. Because it is true, isn't it, that, when we hear of a particle of mass m, we will connect with it a wave field of frequency mc^2/h. And when we encounter a wave field of frequency ν (nu), we link with it energy quanta $h\nu$ or, what is the same thing, mass quanta $h\nu/c^2$. Thus de Broglie's dissertation was the point of departure for the complete uncertainty of our conception of matter. In the image of both particles and waves there are elements of truth which we must not abandon. But we do not know how to combine them.

5. Wave field and particle: their theoretical relationship

The *relationship* of the two images in general is known with great clarity and insuprising detail. No one doubts its correctness and general validity. With regard to the combination into a single, concrete, obvious image, opinions are so divided that many consider it entirely impossible. I will now briefly outline the *relationship*. Do not count on forming a uniform, concrete picture; and do not blame my lack of skill in representation nor your own slowness in understanding for your failure – for up to now no one has been successful.

In a wave, two things are distinguishable: *first* the wave surfaces, which form something like a system of onion skins except that they *spread out* in a direction perpendicular to the skin (i.e., to themselves). The analogy in two (instead of three) dimensions is well known to you in the form of the beautiful circles of waves produced by a stone thrown into a pool. The *second*, less evident, are those imaginary lines perpendicular to the wave surfaces in the direction of which the wave progresses at that point, the *wave normals*, which are also called *rays* – to use for all types of waves an expression familiar in the case of light.

Here I hesitate. For what I want and have to say now is certainly important and fundamental, indeed it is even correct, but in a sense that we shall have to restrict so much that it almost contradicts the provisional assertion. This provisional assertion is: *These wave normals or rays correspond to the particle paths.* If you cut a small piece from the wave, about 10 or 20 waves in the line of propagation, and about as big a piece perpendicular to it and destroy ("calm") the rest of the wave, then such a "wave packet" really moves along a ray with exactly the velocity, or at any rate the change of velocity, that is to be expected from a particle of the type in question at the point in question, having regard to any fields of force present.

Even if, in the wave packet or the wave group, we get a sort of clear picture for the particle, which we can work out in many details (e.g., the *impulse* of the particle increases as the wave length decreases; the two are exactly inversely proportional) – nevertheless, for many reasons, we must not take this clear image quite seriously. In the first place it is really somewhat blurred, and the longer the wave length the more blurred it is. Second, there is often not a small packet but an extended wave. Finally, there can be very small "packets" of such a structure that there can be no question of wave surfaces or wave normals – an important case to which I shall return immediately.

The following interpretation seems to me appropriate and representative, because it is extensively supported by experiment: at each point of a regularly progressing wave train there is a *twofold structural relationship of effects*, which it is possible to differentiate as "longitudinal" and "transverse." The transverse structure is that of the wave surfaces and comes to light in experiments on diffraction and interference; the longitudinal structure is that of wave normals and manifests itself in the observation of single particles. Both have been proved convincingly by ingenious experimental designs, each carefully thought out for its special purpose.

All these concepts of longitudinal and perpendicular structure are not exact and absolute because those of the wave surfaces and wave normals are not. They are necessarily lost when the whole wave phenomenon is limited to a small space by the measurement of a single or very few wave lengths. This case is then of very special interest, above all for those waves that, according to de Broglie, constitute the "second nature" of the electron. For them it turns out that this case must occur in the vicinity of a positively charged atomic nucleus so that the wave phenomenon, a kind of stationary vibration, shrinks into a small space for which the real atomic size is found very exactly by calculation.

This size was in fact already fairly well known in an other manner. Stationary water waves of a similar type can be produced in a small washbasin by splashing fairly regularly in the middle with a finger or even by giving the whole basin a slight jolt, so that the water surface rolls to and fro. Here the waves are no longer regularly distributed; but what attracts attention are the *natural frequencies* of these stationary vibrations, which you also observed quite well in the washbasin. For the wave group acting around the atomic nucleus, we can calculate these frequencies, and they are generally found to be exactly equal to the "energy level" of Bohr's theory (which I mentioned briefly earlier) divided by Planck's constant h. The ingenious but nevertheless somewhat artificial assumptions of that theory, as well as those of the older quantum theory, are replaced by a much more natural assumption in de Broglie's wave phenomena. The wave phenomenon forms the real "body" of the atom. It replaces the individual punctiform electrons, which in Bohr's model swarm around the nucleus. In no case can we admit the existence of such punctiform particles within the atom, and if we still think of the nucleus as such, this is an entirely conscious expedient.

Regarding the discovery that the "energy levels" are really only the frequencies of fundamental vibrations, the point that seems to me particularly important is that we can renounce the postulation of a stepwise transition, because two or more natural vibrations can quite readily be excited simultaneously. The discreteness of the *natural frequencie* suffices completely, as I at least believe, to support the considerations from which Planck started and many similar just as important ones – I mean, in a word, to support the whole of quantum thermodynamics.

6. Quantum steps and identity of particles

The abandonment of the *theory of quantum steps*, which to me personally seems more inadmissible from year to year, has, I admit, important consequences. It really means that the exchange of energy in definitely limited packets is not taken seriously, is not really believed, and is instead replaced by the resonance between vibration frequencies. We have seen, however, that, because of the identity of mass and energy, we must consider the particles themselves as Planckian energy quanta. This is frightening at first. For, with this disbelief, we must also not consider the single particles as a well-defined permanent reality.

There are many other reasons to support the fact that it is not so in reality. In the first place properties have long been attributed to such a particle which contradicts it. From the above fleeting picture of the "wave packet," it is easily possible to infer the famous Heisenberg uncertainty principle, according to which a particle cannot simultaneously be at a specific place and have a clearly determined velocity. Even if this uncertainty were small – and it certainly is not – it follows from it that it is never possible to observe the same particle twice with apodictic certainty.

Another very sound reason to deny to the single particles an identifiable individuality is the following: When we are dealing theoretically with two or more particles of the same type, for example, with the two electrons of a helium atom, we must *efface their identity*, otherwise the results will simply be untrue and not agree with experience. We must count two situations which differ only by an exchange of roles by the electrons not only as equal – that would be obvious – but we must count them as one and the same. If we count them as two equals, it becomes nonsensical.

This situation weighs heavily, because it is valid for every kind of particle in whatever number with out exception and because it is directly contrary to every thing that was believed about them in classic atomic theory.

The fact that the individual particle is not a well-defined permanent object of determinable identity or individuality is probably admitted by most theoreticians just as they admit the reason described here for the complete inadequacy of this representation. Despite this, the single particle still plays a part in their representation, deliberations, discussions and writings, and with this I cannot agree. Even more deeply rooted is the idea of stepwise transition, of "quantum steps," at least according to the words and modes of expression, which have become permanently naturalized – of course in very careful technical language, the ordinary meaning of which is very difficult to grasp.

The term "probability of transition" belongs to the permanent vocabulary, for example. But surely we can speak of the probability of an event only if we think that it sometimes actually happens. And in this case, since we refuse to recognize intermediate states, the transition must of course be a sudden one. If it required time, it might well be interrupted in the middle by an unforeseen disturbance. We would then not know where we were. The allegedly exact and fundamental concept would have a gap. In this concept, moreover, probability plays an overriding part. The delicate dilemma, wave versus particle, should resolve itself in such a manner that from the wave field we could calculate merely the probability of encountering a particle of definite properties at a definite position, if we are looking for such a thing.

This interpretation might be quite in keeping with the findings derived from very high-frequency waves ("ultra-rapid particle streams") with special ingeniously thought-out experimental designs. I mean those that I previously cited as observations on single particles. In the *tracks,* which are called particle paths, a *longitudinal* action relationship along the wave normals undoubtedly comes to light. But such a relationship is altogether to be expected in the propagation of a wave front. In any case we have more chance of understanding it from the wave representation than, contrariwise, recognizing the *transverse* action relationship of interference and diffraction from the combined effects of discrete particles, if the reality of waves is denied and we grant them only a kind of *informative* role.

7. Wave identity

Real existence is a term undoubtedly almost hunted down by many philosophical hounds, and its simple, naïve meaning has almost been lost to us. Therefore, I wish to remind you here of something else. We have spoken of the fact that a particle is not individual. The *same* particles are really never observed twice, just as Heraclitus said of the river. We cannot mark an electron – "paint it red" – and not only that, we cannot even think of them as marked, otherwise false results are obtained by incorrect "counting" step by step – for the structure of line spectra, in thermodynamics, and many other cases.

In contrast, it is very easy to imprint an individual structure on a *wave* so that it can be recognized again with complete certainty. Think of light beacons at sea. Following a definite code, each one has a definite prescribed light sequence, for example, three seconds light, five seconds dark, one second light, again five seconds dark and then light again for three seconds, etc. The sailor knows: That is San Sebastian. Similarly for a whistling buoy; only in this case there are sound waves. Or else: you telephone by radiotelephone to a good friend in New York; as soon as he says, "Hullo, hi, this is Edward Meier," you know that his voice has imprinted a structure on the radio wave which has traveled five thousand miles to you and which can be distinguished with certainty from any other. It is not even necessary to go so far. If your wife calls from the garden, "Frank," it is entirely analogous, only in this case they are merely sound waves. The journey is shorter but takes a little longer. Our whole understanding of language is based on imprinted individual wave structure. And what an abundance of details in rapid succession is transmitted to us by the motion picture or the television picture by following the same principle!

Here we are of course dealing with relatively crude wave structures, with which we should perhaps not compare individual particles but rather the tangible bodies surrounding us. And these have nearly all a very pronounced individuality: my old pocket knife, my old felt hat, Zürich's Münster, etc. I have recognized them with certainty a hundred times. But in a remarkable manner, we find the characteristic that, in contrast to the case of particles, we must ascribe individuality to wave phenomena, in the case of elementary waves.

One example must suffice. A limited volume of helium gas, for example, can be considered to consist of many helium atoms or, *instead*, of a superimposition of elementary wave trains of waves of matter. Both points of view lead to the same results for the behavior of the gas on warming, compression, etc. However, we must proceed differently with certain rather intricate "counts," which have to be undertaken in both cases. If we form a mental picture of the particles, that is, the helium atoms, as I have already stated previously, we must not attribute individuality to them. This originally seemed very surprising and has led to lengthy controversies, which have, however, long since

18

been settled. On the other hand, in the second manner of consideration, which, instead of particles, imagines wave trains of matter, each is accorded a specifiable structure, different from that of any other. To be sure, there exist many pairs, which are so similar to each other that they can exchange their parts, without our noticing it from outside the body of gas. If we tried to *count* the *very many* similar conditions that arise in this way as only one, we would obtain something quite incorrect.

8. Conclusion

You may be surprised that despite everything I finally have put forward, which in reality no one denies, the closely related concepts of the *quantum steps* and the individual particle have not disappeared either from the vocabulary or from the imagination of the physicist. You will find the explanation, if you consider that the interpretation, which we ultimately reached and toward which we steered in about the last third of my lecture of today, invalidates or at least casts doubt on the real significance of many details of the structure of matter that I put forward in the first two-thirds. However, I was able to use – without unbearable verbosity I could not have helped using – a language that I really do not consider appropriate. How can we state the weight of a carbon nucleus and a hydrogen nucleus to an accuracy of several decimal places and establish that the former is slightly lighter than the twelve hydrogen nuclei combined in it, if we do not provisionally accept the point of view that these particles are something quite concretely real? This is so much more convenient and evident that we cannot abandon it, just as the chemist cannot renounce his valence bonds, although he knows quite well that they are a drastic simplification of a quite involved wave-mechanical situation.

If you ask me: Now, really, what *are* these particles, these atoms and molecules? I should have to admit that I know as little about it as where Sancho Panza's second donkey came from. However, to say something, even if not something momentous: They can at the most perhaps be thought of as more or less temporary creations within the wave field, whose structure and structural variety, in the widest sense of the word, are so clearly and sharply determined by means of wave laws as they recur always in the same manner, that much takes place *as if* they were a permanent material reality. The very exactly specifiable mass and charge of the particles we must thus consider as Gestalt elements determined by the wave laws. The conservation of charges and mass on a large scale must be considered as a statistical effect, based on the "law of the large numbers."

WHAT IS AN ELEMENTARY PARTICLE[1]

1. A Particle is not an Individual

Atomism in its latest form is called quantum mechanics. It has extended its range to comprise, besides ordinary matter, all kinds of radiation, including light – in brief, all forms of energy, ordinary matter being one of them. In the present form of the theory the "atoms" are electrons, protons, photons, mesons, etc. The generic name is elementary particle, or merely particle. The term atom has very wisely been retained for chemical atoms, though it has become a misnomer.

This essay deals with the elementary particle, more particularly with a certain feature that this concept has acquired–or rather lost–in quantum mechanics. I mean this: that the elementary particle is not an individual; it cannot be identified, it lacks "sameness." The fact is known to every physicist, but is rarely given any prominence in surveys readable by non-specialists. In technical language it is covered by saying that the particles "obey" a new-fangled statistics, either Einstein-Bose or Fermi-Dirac statistics. The implication, far from obvious, is that the unsuspected epithet "this" is not quite properly applicable to, say, an electron, except with caution, in a restricted

---

[1]This essay was published as Chapter IX in E. Schrödinger, *Science Theory and Man* (George Allen and Unwin Ltd 1957).

21

sense, and sometimes not at all. My objective here is to explain this point and to give it the thought it deserves. In order to create a foil for the discussion, let me summarize in sections 2-5 what we are usually told about particles and waves in the new physics.

2. Current Views: the Amalgamation of Particles and Waves

Our image of the material world had been made up of two kinds of "fittings": waves and particles. The former were instanced mainly, if not exclusively, by Maxwell's waves of electromagnetic energy, comprising such as are used in radio, light, X-rays, and gamma-rays. Material bodies were said to consist of particles. One was also familiar with jets of particles, called corpuscular rays, such as cathode rays, beta-rays, alpha-rays, anode rays, etc. Particles would emit and absorb waves. For instance, cathode rays (electrons), when slowed down by colliding with atoms, emit X-rays. The distinction between particles and waves was, however, considered as clear-cut as that between a violin and its sound. An examinee who alleged cathode rays to be waves, or X-rays to be jets of particles, would have got very bad marks.

In the new setting of ideas the distinction has vanished, because it was discovered that all particles have also wave properties, and *vice versa*. Neither of the two concepts must be discarded, they must be amalgamated. Which aspect obtrudes itself depends not on the physical object, but on the experimental device set up to examine it. A jet of cathode rays, for example, produces in a Wilson cloud chamber discrete tracks of water droplets – curved tracks if there is a magnetic field to deflect the electrons, otherwise straight alignments of droplets. We cannot but interpret them as traces of the paths of single electrons. Yet the same jet, after crossing a narrow tube placed at right angles to it and containing crystal powder, will produce on a photographic plate at some distance behind the tube a pattern of concentric circles. This pattern can be understood in all its details when looked upon as the interference pattern of waves, and in no other way. Indeed, it bears a close resemblance to similarly produced X-ray patterns.

The suspicion arises: are the conical jets that impinge on the photographic plate and form the pattern of circles really cathode rays; are they not perhaps secondary X-rays? The suspicion has to be dismissed, for the whole system of circles can be displaced by a magnet, while X-rays can not; moreover, by putting a lead screen with a small hole in it in the place of the photographic plate, a jetlet can be isolated from one of the conical jets and made to display any of the typical particle characters of cathode rays: it will produce discrete tracks in a cloud chamber; bring about discrete discharges in a Geiger-Müller counter; and charge up a Faraday cage in which it is intercepted. A vast amount of experimental evidence clinches the conviction that wave characteristics and particle characteristics are never encountered singly, but always in a union; they form different aspects of the same phenomenon, and indeed of all physical phenomena. The union is not a loose or superficial one. It would be quite unsatisfactory to consider cathode rays to consist both of particles and of waves. In the early days of the new theory it was suggested that the particles might be singular spots within the waves, actually singularities in the meaning of the mathematician. The white crests on a moderately rough sea would be a fairly adequate simile. The idea was very soon abandoned. It seems that both concepts, that of waves and that of particles, have to be modified considerably, so as to attain a true amalgamation.

23

3. Current Views: the Nature of Waves

The waves, so we are told, must not be regarded as quite real waves. It is true that they produce interference patterns – which is the crucial test that in the case of light had removed all doubts as to the reality of the waves. However, we are now told that all waves, including light, ought rather to be looked upon as "probability waves." They are only a mathematical device for computing the probability of finding a particle in certain conditions, for instance (in the above example), the probability of an electron hitting the photographic plate within a small specified area. There it is registered by acting on a grain of silver bromide. The interference pattern is to be regarded as a statistical registration of the impinging electrons. The waves are in this context sometimes referred to as guiding waves – guiding or directing the particles on their paths. The guidance is not to he regarded as a rigid one; it merely constitutes a probability. The clear-cut pattern is a statistical result, its definiteness being due to the enormous number of particles.

Here I cannot refrain from mentioning an objection which is too obvious not to occur to the reader. Something that influences the physical behaviour of something else must not in any respect be called less real than the something it influences – whatever meaning we may give to the dangerous epithet "real." It is certainly useful to recall at times that all quantitative models or images conceived by the physicist are, epistemologically, only mathematical devices for computing observable events, hut I cannot see that this applies more to, say, light-waves than to, say, oxygen molecules.

4. Current Views: the Nature of Particles (Uncertainty Relation)

As regards the modification required in the concept of a particle, the stress is on Heisenberg's uncertainty relation. The so-called classical mechanics hinged on Galileo's and Newton's discovery that the thing which in a moving body is determined at any instant by the other bodies in its environment is only and precisely its acceleration, or, in mathematical terms, the second derivatives with respect to time of the coordinates. The first derivatives, commonly called the velocity, are therefore to be included in the description of the momentary state of the body, together with the coordinates themselves which label its momentary place in space or "whereness" (or ubiety, to use an antiquated but convenient word). Thus, to describe the momentary state of a particle, two independent data were required: its coordinates and their first time derivatives, or ubiety and velocity. According to the new theory less is required, and less is obtainable. Either of the two data can be given with arbitrary accuracy, provided that no store is set on the other, but both cannot be known together with absolute precision. One may not even conceive of both as having absolutely sharp values at the same instant. They mutually blur each other, as it were. Broadly speaking, the product of the latitudes of their respective inaccuracies cannot be reduced below a fixed constant. For an electron, this constant happens to be about 1 if the units centimetre and second are used. Thus, if the velocity of an electron is considered sharp with a latitude of only 1 cm/sec, its location has to be considered as blurred within the latitude of 1 cm. The strangeness does not lie in the mere existence of inaccuracies, for the particle might he a thing of vague and changeable extension, within which slightly different velocities prevailed at different spots. Then, however, a sharp location or ubiety would probably entail a sharply defined velocity and *vice versa*. Actually it is just the other way round.

5. Current Views: the Meaning of the Uncertainty Relation

Two links connect this strange and certainly very fundamental statement to other parts of the theory. It can be arrived at by declaring that a particle is equivalent to its guiding wave, and has no characteristics save those indicated by the guiding wave according to a certain code. The code is simple enough. The ubiety is indicated by the extension of the wave, the latitude in the velocity by the range of wave numbers. "Wave number" is short for reciprocal of the wavelength. Each wave number corresponds to a certain velocity proportional to it. That is the code. It is a mathematical truism that the smaller a wave group, the wider is the (minimum) spread of its wave numbers.

Alternatively, we may scrutinize the experimental procedure for determining either the ubiety or the velocity. Any such measuring device implies a transfer of energy between the particle and some measuring instrument – eventually the observer himself, who has to take a reading. This means an actual physical interference with the particle. The disturbance cannot be arbitrarily reduced, because energy is not exchanged continuously but in portions. We are given to understand that, when measuring one of the two items, ubiety or velocity, we interfere with the other the more violently the higher the precision we aim at. We blur its value within a latitude inversely proportional to the latitude of error allowed in the first.

In both explanations the wording seems to suggest that the uncertainty or lack of precision refers to the attainable knowledge about a particle rather than to its nature. Indeed, by saying that we disturb or change a measurable physical quantity we logically imply that it has certain values before and after our interference, whether we know them or not. And in the first explanation, involving the wave, if we call it a guiding wave how should it guide the particle on its path, if the particle has not got a path? If we say the wave indicates the probability of finding the particle at A, or at B, or at C ... this seems to imply that the particle is at one, and one only, of these places; and similarly for the velocity. (Actually the wave does indicate both probabilities simultaneously, one by its extension, the other by its wave numbers.) However, the current view does not accept either ubiety or velocity as permanent objective realities. It stresses the word "finding." Finding the particle at point A does not imply that it has been there before. We are more or less given to understand that our measuring device has brought it there or "concentrated" it at that point, while at the same time we have disturbed its velocity. And this does not imply that the velocity "had" a value. We have only disturbed or changed the probability of finding this or that value of the velocity if we measure it. The implications as to "being" or "having" are misconceptions, to be blamed on language. Positivist philosophy is invoked to tell us that we must not distinguish between the knowledge we can obtain of a physical object and its actual state. The two are one.

6. Criticism of the Uncertainty Relation

I will not discuss here that tenet of positivist philosophy. I fully agree that the uncertainty relation has nothing to do with incomplete knowledge. It does reduce the amount of information attainable about a particle as compared with views held previously. The conclusion is that these views were wrong and we must give them up. We must not believe that the completer description they demanded about what is really going on in the physical world is conceivable, but in practice unobtainable. This would mean clinging to the old view. Still, it does not necessarily follow that we must give up speaking and thinking in terms of what is really going on in the physical world. It has become a convenient habit to picture it as a reality. In everyday life we all follow this habit, even those philosophers who opposed it theoretically, such as Bishop Berkeley. Such theoretical controversy is on a different plane. Physics has nothing to do with it. Physics takes its start from everyday experience, which it continues by more subtle means. It remains akin to it, does not transcend it generically, it cannot enter into another realm. Discoveries in physics cannot in themselves – so I believe – have the authority of forcing us to put an end to the habit of picturing the physical world as a reality.

I believe the situation is this. We have taken over from previous theory the idea of a particle and all the technical language concerning it. This idea is inadequate. It constantly drives our mind to ask for information which has obviously no significance. Its imaginative structure exhibits features which are alien to the real particle. An adequate picture must not trouble us with this disquieting urge; it must be incapable of picturing more than there is; it must refuse any further addition. Most people seem to think that no such picture can be found. One may, of course, point to the circumstantial evidence (which I am sorry to say is not changed by this essay) that in fact none has been found. I can, however, think of some reasons for this, apart from the genuine intricacy of the case. The palliative, taken from positivist philosophy and purporting to be a reasonable way out, was administered fairly early and authoritatively. It seemed to relieve us from the search for what I should call real understanding; it even rendered the endeavour suspect, as betraying an unphilosophical mind – the mind of a child who regretted the loss of its favourite toy (the picture or model) and would not realize that it was gone for ever. As a second point, I submit that the difficulty may be intimately connected with the principal subject of this paper, to which I shall now turn without further delay. The uncertainty relation refers to the particle. The particle, as we shall see, is not an identifiable individual. It may indeed well be that no individual entity can be conceived which would answer the requirements of the adequate picture stated above.

It is not at all easy to realize this lack of individuality and to find words for it. A symptom is that the probability interpretation, unless it is expressed in the most highly technical language of mathematics, seems to be vague as to whether the wave gives information about one particle or about an *ensemble* of particles. It is not always quite clear whether it indicates the probability of finding "the" particle or of finding "a" particle, or indicates the likely or average number of particles in, say, a given small volume. Moreover the most popular view on probability tends to obliterate these differences. It is true that exact mathematical tools are available to distinguish between them. A point of general interest is involved, which I will explain. A method of dealing with the problem of many particles was indicated in 1926 by the present writer. The method uses waves in many-dimensional space, in a manifold of $3N$ dimensions, N being the number of particles. Deeper insight led to its improvement. The step leading to this improvement is of momentous significance. The many-dimensional treatment has been superseded by so-called second quantization, which is mathematically equivalent to uniting into one three-dimensional formulation the cases $N = 0, 1, 2, 3 \ldots$ (to infinity) of the many-dimensional treatment. This highly ingenious device includes the so-called new statistics, with which we shall have to deal below in much simpler terms. It is the only precise formulation of the views now held, and the one that is always used. What is so very significant in our present context is that one cannot avoid leaving indeterminate the number of the particles dealt with. It is thus obvious that they are not individuals.

7. The Notion of a Piece of Matter

I wish to set forth a view on matter and the material universe, to which Ernst Mach [1], Bertrand Russell [2], and others were led by a careful analysis of concepts. It differs from the popular view. We are, however, not concerned with the psychological origin of the concept of matter but with its epistemological analysis. The attitude is so simple that it can hardly claim complete novelty; some pre-Socratics, including the materialist Democritus [3], were nearer to it than were the great men who resuscitated science and moulded it during the seventeenth to nineteenth centuries.

According to this view, a piece of matter is the name we give to a continuous string of events that succeed each other in time, immediately successive ones being as a rule closely similar. The single event is an inextricable complex of sensates, of associated memory images, and of expectations associated with the former two. The sensates prevail in the case of an unknown object, say a distant white patch on the road, which might be a stone, snow, salt, a cat or a dog, a white shirt or blouse, a handkerchief. Even so, within the ensuing string of events we usually know from general experience how to discount the changes caused by motions of our own body, in particular of our direction of sight. As soon as the nature of the object is recognized, images and expectations begin to prevail. The latter concern sensations as hard, soft, heavy, flexible, rough, smooth, cold, salty, etc., associated with the image of touching and handling; they also concern spontaneous movements or noises such as barking, mewing, shouting, etc. It should be noted that I am not speaking of our thoughts or considerations about the object, but of what forms part and parcel of our perception of it – of what it is to us. However, the limit is not sharp. As our familiarity with a piece of matter grows, and in particular as we approach its scientific aspect, the range of expectations in regard to it widens, eventually to include all the information science has ascertained, e.g. melting point, solubility, electric conductivity, density, chemical and crystalline structure, and so on. At the same time, the momentary sensational core recedes in relevance the more the object becomes familiar to us, whether by scientific knowledge or by everyday use.

8. Individuality or 'Sameness'

After a certain wealth of association has come to outshine the core of sensates, the latter is no longer needed to keep the complex together. It persists even when the contact of our senses with the object temporarily ceases. And more than that: the complex is latently conserved even when the whole string is interrupted by our turning away from the object to others and forgetting all about it. Indeed, this is not exceptional, but a rule which – since we sometimes sleep – has no exception. But we have adopted the useful device of filling these gaps. We supplement the missing parts of the strings relating to pieces of matter in our nearer and farther surroundings, to cover the periods when we neither watch them nor think of them. When a familiar object re-enters our ken, it is usually recognized as a continuation of previous appearances, as being the same thing. The relative permanence of individual pieces of matter is the most momentous feature of both everyday life and scientific experience. If a familiar article, say an earthenware jug, disappears from your room, you are quite sure somebody must have taken it away. If after a time it reappears, you may doubt whether it really is the same one – breakable objects in such circumstances are often not. You may not be able to decide the issue, but you will have no doubt that the doubtful sameness has an indisputable meaning – that there is an unambiguous answer to your query. So firm is our belief in the continuity of the unobserved parts of the strings!

No doubt the notion of individuality of pieces of matter dates from time immemorial. I suppose animals must have it in some way, and a dog, when seeking for his ball that has been hidden, displays it very plainly. Science has taken it over as a matter of course. It has refined it so as safely to embrace all cases of apparent disappearance of matter. The idea that a log which burns away first turns into fire, then into ashes and smoke, is not alien to the primitive mind. Science has substantiated it; though the appearance in bulk may change, the ultimate constituents of the matter do not. This was (in spite of his occasional scepticism mentioned above) the teaching of Democritus. Neither he nor Dalton doubted that an atom which was originally present in the block of wood is afterwards either in the ashes or in the smoke.

9. The Bearing on Atomism

In the new turn of atomism that began with the papers of Heisenberg and of de Broglie in 1925 such an attitude has to be abandoned. This is the most startling revelation emerging from the ensuing development, and the feature which in the long run is bound to have the most important consequences. If we wish to retain atomism we are forced by observed facts to deny the ultimate constituents of matter the character of identifiable individuals. Up to recently, atomists of all ages, for all I know, had transferred that characteristic from visible and palpable pieces of matter to the atoms, which they could not see or touch or observe singly. Now we do observe single particles; we see their tracks in the cloud chamber and in photographic emulsions; we register the practically simultaneous discharges caused by a single swift particle in two or three Geiger counters placed at several yards' distance from each other. Yet we must deny the particle the dignity of being an absolutely identifiable individual. Formerly, if a physicist were asked what stuff the atoms themselves were made of, he might smile and shirk the answer. If the inquirer insisted on the question whether he might imagine them as small unchangeable bits of ordinary matter, he would get the smiling reply that there was no point in doing so but that it would do no harm. The formerly meaningless question has now gained significance. The answer is definitely in the negative. An atom lacks the most primitive property we associate with a piece of matter in ordinary life. Some philosophers of the past, if the case could be put to them, would say that the modern atom consists of no stuff at all but is pure shape.

10. The Meaning of the New Statistics

We must at last proceed to give the reasons for this change of attitude in a more comprehensible form than at the end of section 6. It rests on the so-called new statistics. There are two of them. One is the Bose-Einstein statistics, whose novelty and relevance were first stressed by Einstein. The other is the Fermi-Dirac statistics, of which the most pregnant expression is Pauli's exclusion principle. I shall try to explain the new statistics, and its relation to the old classical or Boltzmann statistics, to those who have never heard about such things and perhaps may be puzzled by what "statistics" means in the context. I shall use an instance from everyday life. It may seem childishly simple, particularly because we have to choose small numbers – actually 2 and 3 – in order to make the arithmetic surveyable. Apart from this, the illustration is completely adequate and covers the actual situation.

Three schoolboys, Tom, Dick, and Harry, deserve a reward. The teacher has two rewards to distribute among them. Before doing so, he wishes to realize for himself how many different distributions are at all possible. This is the only question we investigate (we are not interested in his eventual decision). It is a statistical question: to count the number of different distributions. The point is that the answer depends on the nature of the rewards. Three different kinds of reward will illustrate the three kinds of statistics.

(a) The two rewards are two memorial coins with portraits of Newton and Shakespeare respectively. The teacher may give Newton either to Tom or to Dick or to Harry, and Shakespeare either to Tom or to Dick or to Harry. Thus there are three times three, that is nine, different distributions (classical statistics).

(b) The two rewards are two shilling-pieces (which, for our purpose, we must regard as indivisible quantities). They can be given to two different boys, the third going without. In addition to these three possibilities there are three more: either Tom or Dick or Harry receives two shillings. Thus there are six different distributions (Bose-Einstein statistics).

(c) The two rewards are two vacancies in the football team that is to play for the school. In this case two boys can join the team, and one of the three is left out. Thus there are three different distributions (Fermi-Dirac statistics).

Let me mention right away the *rewards* represent the particles, two of the same kind in every case; the boys represent states the particle can assume. Thus, "Newton is given to Dick" means: the particle Newton takes on the state Dick.

Notice that the counting is natural, logical, and indisputable in every case. It is uniquely determined by the nature of the objects: memorial coins, shillings, memberships. They are of different categories. Memorial coins are individuals distinguished from one another. Shillings, for all intents and purposes, are not, but they are still capable of being owned in the plural. It makes a difference whether you have one shilling, or two, or three. There is no point in two boys exchanging their shillings. It does change the situation, however, if one boy gives up his shilling to another. With memberships, neither has a meaning. You can either belong to a team or not. You cannot belong to it twice over.

Experimental evidence proves that statistical counts referring to elementary particles must never follow the pattern (a), but must follow either (b) or (c). Some hold that for all genuinely elementary particles (c) is competent. Such particles, electrons for instance, correspond to membership in a club; I mean to the abstract notion of membership, not to the members. Any person eligible to membership in that club represents a well-defined state an electron can take on. If the person is a member, that means there is an electron in that particular state. According to Pauli's exclusion principle, there can never be more than one electron in a particular state. Our simile renders this by declaring double membership meaningless – as in most clubs it would be. In the course of time the list of members changes, and membership is now attached to other persons: the electrons have gone over into other states. Whether you can, in a loose way, speak of a certain membership going over from Dick to Tom, thence from Tom to Harry, etc., depends on the circumstances. They may suggest this view, or they may not, but never in an absolute fashion. In this our simile is perfect, for it is the same with an electron. Moreover, it is quite appropriate to consider the number of members as fluctuating. Indeed, electrons too are created and annihilated.

The example may seem odd and inverted. One might think, "Why cannot the people be the electrons and various clubs their states? That would be so much more natural." The physicist regrets, but he cannot oblige. And this is just the salient point: the actual statistical behaviour of electrons cannot be illustrated by any simile that represents them by identifiable things. That is why it follows from their actual statistical behaviour that they are not identifiable things.

The case (b), illustrating Einstein-Bose statistics, is competent for light quanta (photons), *inter alia*. It hardly needs discussion. It does not strike us as so strange for the very reason that it includes light, i.e. electro-magnetic energy; and energy, in prequantum times, had always been thought of in very much the way our simile represents it, viz. as having quantity, but no individuality.

11. Restricted Notion of Identity

The most delicate question is that of the states of, say, an electron. They are, of course, to be defined not classically, but in the light of the uncertainty relation. The rigorous treatment referred to at the end of section 6 is not really based on the notion of "state of one electron" but on that of "state of the assembly of electrons." The whole list of members of the club, as it were, has to be envisaged together – or rather several membership lists, corresponding to the several kinds of particles that go to compose the physical system under consideration. I mention this, not to go into details about it, but because, taken rigorously, the club simile has two flaws. First, the possible states of an electron (which we had assimilated to the persons eligible for membership) are not absolutely defined; they depend on the arrangement of the – actual or imagined – experiment. Given this arrangement, the states are well-defined individuals, which the electrons are not. They also form – and this is the second flaw of the simile – a well-ordered manifold. That is, there is a meaning in speaking of neighbouring states as against such as are farther remote from each other. Moreover, I believe it is true to say that this order can be conceived in such a fashion that, as a rule, whenever one occupied state ceases to be occupied, a neighbouring state becomes occupied.

This explains that, in favourable circumstances, long strings of successively occupied states may he produced, similar to those contemplated in sections 7 and 8. Such a string gives the impression of an identifiable individual, just as in the case of any object in our daily surrounding. It is in this way that we must look upon the tracks in the cloud chamber or in a photographic emulsion, and on the (practically) simultaneous discharges of Geiger counters set in a line, which discharges we say are caused by the same particle passing one counter after another. In such cases it would be extremely inconvenient to discard this terminology. There is, indeed, no reason to ban it, provided we are aware that, on sober experimental grounds, the sameness of a particle is not an absolute concept. It has only a restricted significance and breaks down completely in some cases.

In what circumstances this restricted sameness will manifest itself is fairly obvious; namely, when only few states are occupied in the region of the state-manifold with which we are concerned, or, in other words, when the occupied states are not too crowded in that region, or when occupation is a rare event the terms few, crowded, and rare all referring to the state-manifold. Otherwise, the strings intermingle inextricably and reveal the true situation. In the last section we shall formulate the quantitative condition for the prevailing of restricted individuality. Now we ask what happens when it is obliterated.

12. Crowdedness and Wave Aspect

One gains the impression that according as the individuality of the particles is wiped out by crowding, the particle aspect becomes altogether less and less expedient and has to be replaced by the wave aspect. For instance, in the electronic shell of an atom or molecule the crowding is extreme, almost all the states within a certain region being occupied by electrons. The same holds for the so-called free electrons inside a metal. Indeed, in both cases the particle aspect becomes entirely incompetent. On the other hand, in an ordinary gas the molecules are extremely rare in the wide region of states over which they spread. No more than one state in 10,000 or so is occupied. And, indeed, the theory of gases, based on the particle aspect, was able to attain great perfection long before the wave nature of ordinary matter was discovered. (In the last remark I have been speaking of the molecules as if they were ultimate particles; this is legitimate as far as their translatory motion is concerned.)

It is tempting to assign to the two rivals, the particle aspect and the wave aspect, full competences in the limiting cases of extreme "rarefaction" and extreme "crowding" respectively. This would separate them, as it were, with only some sort of transition required for the intermediate region. This idea is not entirely wrong, hut it is also far from correct. One may remember the interference patterns referred to in section 2 in evidence of the wave nature of the electron. They can be obtained with an arbitrarily faint bundle of cathode rays, provided the exposure is prolonged. Thus a typical wave phenomenon is produced here, irrespective of crowding. Another instance is this. A competent theoretical investigation of the collision of two particles, whether of the same or of different kind, has to take account of their wave nature. The results are duly applied to the collisions of cosmic ray particles with atomic nuclei in the atmosphere, both being extremely rarefied in every sense of the word. But perhaps this is trivial; it only means that even an isolated particle, which gives us the illusion of transitory individuality, must yet not be likened to a classical particle. It remains subject to the uncertainty relation, of which the only tolerable image is the guiding wave group.

13. The Condition for the Particle Aspect

The following is the quantitative condition for strings to develop which counterfeit individuals and suggest the particle aspect: the product of the momentum p and the average distance l between neighbouring particles must be fairly large compared with Planck's constant h; thus

$$pl \gg h.$$

(The momentum p – and not the velocity – is the thing we should really have referred to when, in sections 4 and 5, we dealt with the uncertainty relation; p is simply the product of the mass and the velocity, unless the latter is comparable with that of light.)

A large l means a low density in ordinary space. What matters, however, is the density in the manifold of states – or phase space, to use the technical term. That is why the momentum p comes in. It is gratifying to remember that those very obvious strings – visible tracks in the cloud chamber or in the photographic emulsion, and simultaneous discharges of aligned counters – are all produced by particles with comparatively very large momentum.

The above relation is familiar from the theory of gases, where it expresses the condition which must be fulfilled in very good approximation in order that the old classical particle theory of gases should apply in very good approximation. This theory has to he modified according to quantum theory when the temperature is very low and at the same time the density very high, so that the product pl is no longer very large compared with h. This modification is called the theory of degenerate gases, of which the most famous application is that by A. Sommerfeld to the electrons inside a metal; we have mentioned them before as an instance of extreme crowding.

There is the following connection between our relation and the uncertainty relation. The latter allows one at any moment to distinguish a particle from its neighbours by locating it with an error considerably smaller than the average distance l. But this entails an uncertainty in p. On account of it, as the particle moves on, the uncertainty in the location grows. If one demands that it still remain well below l after the particle has covered the distance l, one arrives precisely at the above relation.

But again I must warn of a misconception which the preceding sentences might suggest, viz. that crowding only prevents us from registering the identity of a particle, and that we mistake one particle for the other. The point is that they are not individuals which could be confused or mistaken one for another. Such statements are meaningless.

REFERENCES

[1] Mach, Ernst. *Erkenntnis und Irrtum*, J. A. Barth, Leipzig. 1905.

[2] Russell, Bertrand. *Human Knowledge, Its Scope and Limits*, Allen and Unwin, London. 1948.

[3] Diels, Hermann. *Die Fragmente der Vorsokratiker*, Weidmannsche Buchhandlung, Berlin. 1903. (The reference is mainly to the fragment 125 of Democritus.)

Expanding Universes[*]

[*]Original publication: Erwin Schroödinger, *Expanding Universes*
(Cambridge University Press, Cambridge 1956)

PREFACE

This brief course of lectures, delivered in the summer term 1954 to an advanced seminar group, does not and cannot exhaust the subject. The general investigation of the matter tensor that would support any particular form of expanding universe is entirely left aside; the line element is regarded as given, and the main objective is the behaviour of test particles and light-signals, the results obtained by observing them and the inferences drawn therefrom.

The de Sitter universe is dealt with at great length. On account of the fact that its matter tensor vanishes, this universe allows of several equally simple representations, which are so different that one is rather amazed at their representing the same geometrical object. Regarding as the basic form of this object the well-known one-shell (hyper-)hyperboloid, I have tried to make visualizable the relationship between the contracting and expanding (hyper-)spherical frame, the static frame, and the expanding flat frame. Each of them is characterized by the bundle of (hyper-)planes whose intersections with the (hyper-)hyperboloid yield the family of "contemporary" spaces, i.e. space for constant time. Of particular interest, is the fact that the second and third frames do not use the whole basic object, i.e. the whole (hyper-)hyperboloid, for representing the whole world; the static frame confines the world to a comparatively small section of the basic object, the flat frame to just half of it. This entails the possibility, already noticed by A. S. Eddington, that extra-mundane test particles and light-signals can enter the ken of an observer, situated inside those respective sections, and cause him intellectual worries, if he tries to interpret them in one of these two frames. fn the static frame he must, even realize the possibility that he himself may be "catapulted" into regions outside his world. This would not even be prevented by admitting the so-called elliptic interpretation of the basic object (i.e. identifying its antipodal points), a possibility to which some thought is devoted in an earlier section.

In Chapter II, I consider the behaviour of test particles and light-signals in more general expanding universes assuming that they are time-like and null geodesics, respectively; the main features are the gradual spending of momentum (or kinetic energy) by a test particle and the analogous reddening of light-signals. The connection of these phenomena with the "work done against an imaginary receding piston" is pointed out, and an important remark of R. C. Tolman is stressed, viz. that an independent proof that our actual universe is certainly *not* static is afforded by the changes in, *rest* mass which are not only observed in the laboratory, but are well ascertained to take place permanently on a large scale in the interior of the stars.

In Chapters III and IV, I endeavour to show that the assumption about the paths being geodesics is well supported by the wave theories of light and matter. The wave theory of light also supplies proof that a (nearly) homogeneous parcel of light decreases its total energy content proportionally to its frequency, while its linear dimensions, if the spread by diffraction is disregarded, increase along with, and proportionally to, the radius of the universe, as does the wave-length. In principle the same statements hold for matter waves, but are of less immediate application; moreover, they are rendered vaguer on account of the longitudinal spread caused by the dispersion.

My warm thanks are due to Mr Alfred Schulhof for the meticulous care which he devoted to constructing exact parallel perspective views of the reduced basic object representing the de Sitter universe, and to tracing the families of plane sections, characteristic of the various ways (or frames of reference) in which this interesting solution of the cosmological field equations can be conceived. I trust that the understanding of their mutual relationship will be facilitated by these accurate drawings.

Dublin 1955   E.S.

I. THE DE SITTER UNIVERSE

1. Synthetic construction

The simplest way of obtaining a model of the de Sitter universe is to proceed by synthesis as follows. Let x, u, v, y, z be the Cartesian coordinates in a Euclidean R_5. Envisage the hypersphere

$$x^2 + u^2 + v^2 + y^2 + z^2 = R^2. \tag{1}$$

All its points are of eqrial right (equivalent). At every point all directions are equivalent. Hence its intrinsic metrical tensor g_{ik} is virtually the same everywhere, and at any point enjoys full spherical symmetry. The same holds for its conracted curvature tensor R_{ik}. Hence these two tensors must be proportional

$$R_{ik} = \Lambda g_{ik}, \tag{2}$$

with Λ a constant, depending only on the radius R. (Actually $\Lambda = -3/R^2$.) The group of linear transformations of the coordinates that leaves (1) unchanged is the ten-parameter group of rotations[1] around the origin in R_5.

If instead of (1) we envisage the one-shell (hyper-) hyperboloid

$$x^2 + u^2 + v^2 + y^2 - z^2 = R^2, \tag{3}$$

---

[1]Including reflections.

49

then *on   it* (2) will again hold, provided that we now understand the g_{ik} *on   it* to mean the intrinsic metric induced by that pseudo-Euclidean geometry of the R_5 that defines the square of "distance" between any two points (x_1, u_1, etc. and x_2, u_2, etc.) by

$$(x_1 - x_2)^2 + (u_1 - u_2)^2 + (v_1 - v_2)^2 + (y_1 - y_2)^2 - (z_1 - z_2)^2, \tag{4}$$

and understand, of course, the R_{ik} to be formed from *these* g_{ik}. (This is clear by pure algebra, since the present case results formally from the previous one, on replacing z by iz.) However, to conform with convention, we prefer to change the sign in the definition (4) (but we do not change anything in (3)!). This changes the sign of ten g_{ik} in (2), but not of the R_{ik}, which are homogeneous, of degree zero, in the g_{ik}. Hence the only thing that changes is the relation between Λ and R^2. (Now $\Lambda = +3/R^2$.) The linear automorphisms of (3) are those induced by the pseudo-rotations (including reflections) in our R_5 around the origin; they contain both the wider and the narrower "Lorentz group in five dimensions." We shall mainly use the narrower, i.e. the one with $\partial z/\partial z'$ positive. It is, of course, ten-parametric as in the previous case. On our four-dimensional hyper-surface (3), *which is our model of the de Sitter universe*, it plays the role that in Minkowski space is played by the (usually so-called) six-parameter Lorentz group, enhanced by the four-parameter translation group. Actually our present ten-parameter group allows us to shift every point on (3) into every other one, showing that they are all equivalent just as the points on (1). The de Sitter space-time is completely homogeneous. (The shift is by no means restricted to a conservation of the sign of z, even if we demand that $\partial z/\partial z' > 0$; the reason is that all points on (3) are situated in a space-like direction with respect to the origin in R_5.)

There is still another question: Will the metric induced on (3) have the required signature? In R_5 it, is four negative, one positive sign. Is it, on the R_4,represented by (3), the Minkowski one? (Three minus, one plus.) This is not a matter of course. But we have from (3), for dx, etc., a line element on (3),

$$+xdx + udu + vdv + ydy - zdz = 0.$$

Thus all these line elements are orthogonal to the radius vector, which is space-like. So you may at any point of (3) choose a local orthogonal frame in five dimensions, one of the axes being in the direction of the radius vector. This one takes one of the negative signs, and this one drops out, for the line elements in the R_4. The remaining four represent a Minkowski metric.

Let us still observe – it will interest us later – that the close relationship between (1) and (3) affords an easy determination of the geodesics on (3). They must correspond to those on (1), i.e. to the great circles on the hypersphere. Now the latter are cut out by all planes through the origin in the Euclidean R_5. To them correspond – since the algebraic change $z \to iz$ is linear and homogeneous – planes through the origin in the pseudo-Euclidean R_5. So all the geodesics on (3) – geodesics on the surface, not in the R_5 – are cut out by the planes through the origin of the R_5, and conversely. They are thus plane curves, indeed conic sections. The space-like are ellipses, the time-like are hyperbola branches. These things will become clearer in the reduced model.

2. The reduced model. Geodesics

To obtain a visualizable model, we shall now suppress the coordinates u and v or, better, we fix our attention on the on the *cross-section*

$$u = 0, v = 0 \qquad (5)$$

of the full model. The embedding R_5 becomes thus an R_3
with Minkowski metric $(ds^2 = -dx^2 - dy^2 + dz^2)$, which
can in the familiar way be visualized in our space of perception. The universe (3) is reduced to an ordinary one-shell
equilateral hyperboloid

$$x^2 + y^2 - z^2 = R^2, \qquad (H)$$

leaving for *space* just one dimension. This is regrettable,
for it will not allow us to answer some questions directly,
e.g. what is the *general* spatial shape of a geodesic orbit.
(In *one*-dimensional *space* we can only find straight lines –
geodesics, all right, but probably not the general type.)
But for many other questions the reduced and thereby
visualizable model is useful. Let us observe, by the way,
that the reduction by *two* dimensions – thus $R_5 \to R_3$,
for the embedding manifold and $R_3 \to R_1$ for space – is in
some respects less misleading than would be a reduction by
only one step (which would not vouchsafe visualizability
anyhow). It conserves the *parity*. In R_3, just as in R_5,
every Lorentz transformation has necessarily an invariant
axis, in R_4 has not. Moreover, *wave propagation* in one-
dimensional *space* is of the same general type as in R_3.
In *spaces* with an even number of dimensions it is not
(*Nachhall!*).

Let me emphasize again, that the metric on (H) is
precisely the one that results from accepting Minkowski
metric in the embedding R_3, with z playing the role of
"time":

$$ds^2 = -dx^2 - dy^2 + dz^2, \qquad (6)$$

and that Lorentz transformations around the origin in R_3
transform (H) into itself and leave the aforesaid metric

52

on (H) untouched. (In (6) we could eliminate one of the three differentials and the corresponding coordinate, from equation (H), but for the moment there is no point in doing so.)

We shall now interpret, z as the world time. This is,of course, not, necessary; later on we shall contemplate other choices. It, must, not, be forgotten, by the way that from the point of view of general relativity, *any* changes of co-ordinates on (H) are admissible, not, only the aforesaid automorphisms of (H). But they do form a very distinguished set.

With z taken as time, the parallel circles on (H) represent space at different, times. Thus the circumference of space (measured according to (6), with $dz = O$) contracts up to a certain epoch, $z = 0$, and then expands. This epoch and the events on the "bottle-neck" appear to be distinguished. This cannot be, since we know that all points on (H) are equivalent. Are then all parallels equivalent? Not at all. We have just stated that they have different invariant circumferences! Moreover, the bottle-neck parallel is a (spatial) geodesic, the others are not. (Indeed, the former is cut out by a plane through the origin in our reduced model; and this means also a plane through the origin of the full model, since the equations (5) are linear homogeneous). But a Lorentz transformation of R_3, involving also z, turns the bottle-neck parallel into an ellipse cut out of (H) by some plane through the origin, a plane subtending an angle less than $45°$ with the xy-plane. Hence all these ellipses are equivalent. In their entirety they represent the entirety of spatial geodesics that show up in our reduced model. *Any one of them* is in a particular frame the bottle-neck! And *any* point of (H) can be transferred into the bottle-neck by a suitable Lorentz transformation of R_3 We shall soon have to pay special attention to antipodic pairs of points on (H). We call the antipode of

53

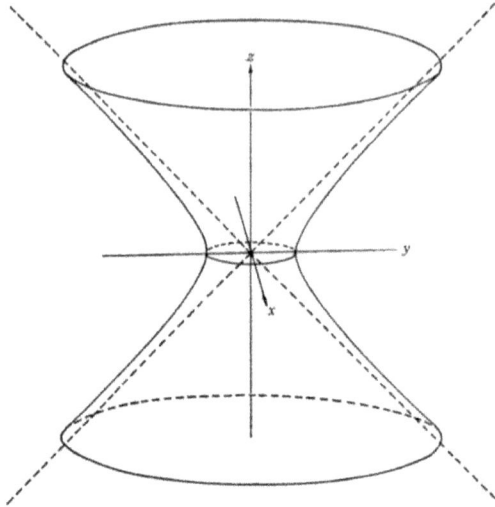

Fig. 1 The basic geometrical object that represents the de Sitter universe, reduced from five to three dimensions. Exhibited are the directions of the x-, y- and z-axes drawn from the origin O of the embedding Euclidean (or rather Minkowskian) R_3. The three ellipses are examples of "contemporary spaces" if z is taken to be the time. The small ellipse is called the "bottleneck", the two large ones are adopted as artificial boundaries for representing in our figure the actually infinite hyperboloid by just its middle part. The profile hyperbola very nearly coincides with a meridian hyperbola. The intersection of the hyperboloid with any plane passing through O is a geodesic, time-like when it is a hyperbola branch, space-like when it is an ellipse, null geodesic in the degenerate limiting case, when it is straight.

$P(x, y, z)$ the point $\bar{P}(-x, -y, -z)$. We make a note of the obvious fact that, if a Lorentz transformation carries P into Q, it, carries $\bar{P}$ into $\bar{Q}$. The relation of antipodicity is Lorentz invariant.

Envisage one of the aforesaid planes through the origin, at an angle less than $45°$ with the xy-plane. It cuts out a spatial geodesic, an ellipse. This becomes more and more elongated as you let the angle approach to $45°$. When the angle reaches 45°, the ellipse degenerates into a couple of parallel *generators*, each of which is the locus of antipodes of the points on the other one. One belongs to the one *family* of generators, the other to the other. Call those two generators g_1 and g_2. Notice that g_2 is the only generator of the second family which has no finite point of intersection with g_1; notice also that two generators of the same family never intersect.

From our construction we know the generators to be geodesics. Since they make an angle of 45° with the xy-plane they are, from (6), null geodesics (light-rays). By further increasing the slope of the plane from 45° to 90°? we get the time-like geodesics (hyperbolae), but these do not interest us at the moment. The entirety of generators of both families form the entirety of null geodesics in our reduced model. Is this property of the null geodesics, being straight lines, only a consequence of our simplification of the model? No. It holds also for the full model. For a null *line element* on the hyper-surface (3) must, of course, be one also in the R_5. Thus it must subtend an angle of 45° with the z-axis. Therefore a null geodesic on (3) must have a constant angle of 45° with the z-axis all along its course. Now we have seen that, all geodesics are plane curves, thus also the null geodesic. But, a *plane* curve making a constant, angle with a given direction is necessarily straight. So we see that this is an essential property, not brought about by the reduction. However, we now return to the reduced model. Through any given point on (H) there is one of each family, the two forming the "light-cone" at that point. They are apparently (i.e. Euclideanly) orthogonal for points at the bottle-neck. For points on bigger parallels, either way: the inner angle of their "light-cone" closes apparently, approaching to zero at infinity. The light-cone at any point P can, according to what we explained before, be obtained as the intersection of (H) with the couple of

45°-planes passing through P and the origin O, the *total* intersection consisting obviously of both the "light-cone" at, P and the one at its antipode $\bar{P}$. The "light-cone" at P alone can also be obtained by intersecting (H) with the tangential plane at P; or alternatively by intersecting (H) with that (two-dimensional) light-cone of the embedding R_3 that has its apex at P. (Henceforth the word *light-cone* shall refer to the couple of null geodesics (light-rays) on (H), unless the contrary is expressly stated.)

Notice in particular the following. The interiors of the light-cones of antipodes P and $\bar{P}$ have no finite point in common, since neither of the two rays through P can cut either of the two through $\bar{P}$, partly for being parallel, partly for belonging to the same family. With non-antipodes at least one, usually two, intersections of the two light-cones occur; for the antipode $\bar{P}$ of P is uniquely defined by having its light-cone parallel to that of P. Hence non-antipodes always have parts of the interiors of their light-cones in common.

3. The elliptic interpretation

We wish to discuss in our reduced model the so-called "elliptic" interpretation of the pseudosphere (H): whether it is feasible, what recommends it and what are its consequences. It, consists in deeming antipodes to represent the same world-point or event. In this discussion we shall speak of our reduced model as if it were the full model, deferring to a separate consideration whether the inferences drawn from the former apply to the latter. *Moreover, we shall not at first introduce the elliptic interpretation.* For this might mean begging the question; at any rate it would confuse the issue. We shall first use the naïve interpretation and prove theorems that have a bearing on the more sophisticated one.

If P has a large positive z, $\bar{P}$ has a large negative z. This raises the fear that an observer might witness in the distant future the same event again – which he has once witnessed in the remote past – if P and $\bar{P}$ meant the same event. However, this fear is unfounded, since we have seen that antipodes are joined by a space-like geodesic, so that an observer cannot possibly move from one to the other.

But it would be equally awkward (for the intended identification of P and $\bar{P}$) if an observer could at some epoch of his life obtain, by a signal, information about P, and at, some later epoch meet, with a signal from $\bar{P}$. If that happened he would certainly deem P and $\bar{P}$ to be different events. But this also cannot happen. For in order to receive a signal from P, he must enter the after-cone (the cone of the future of P). Having got into it, he can never get out again. (For it is impossible to cross an after-cone from inside to outside without surpassing the velocity of light – an after-cone is the most efficient jail in the world.) Now we have seen that the light-cones of P and $\bar{P}$ have no point in common. Hence our observer can never again enter either the after-cone or the fore-cone (the cone of the past) of P. This double statement is valuable. Indeed, the second eventuality would enable our observer, after receiving a message from P, to send a message to $\bar{P}$, which would be still more disastrous for the intended identification. But when I just said that the double statement was valuable, I also meant this: if later on we actually adopt this identification, we shall probably have to exchange the notions of fore-cone and after-cone for $\bar{P}$. But for the moment we do not adopt it, and we continue to regard as after-cone that, half that opens towards the positive z.

Perhaps it is useful to meet a possible objection. It might be thought that the theorems of the preceding paragraph could be outwitted by a relay of two or more observers giving messages to each other. Could they perhaps succeed in uniting in the mind of one of them a message from P and a message from $\bar{P}$? No. For a relay of observers is nothing but a complicated signal. No such arrangement can ever smuggle a message picked up from P outside the after-cone of P, and the same holds in the case of $\bar{P}$. These two regions exclude each other, so the two messages cannot meet. And for the same reason no such arrangement can smuggle a message from P into the fore-cone of $\bar{P}$.

This removes the *prima facie objections* to the intended, identification of antipodes. But there is a circumstance which, to my view, positively suggests it. It is this. The potential experience of an observer who moves on an arbitrary time-like world line from the infinite past $(z \to -\infty)$ to the infinite future $(z \to \infty)$ includes just one half of the surface (H). Which half depends on the world line he follows. But, according to what we know already, it, certainly cannot contain more than one only out of every couple of antipodic points. In other words, if our statement, is correct, the inaccessible half consists of the antipodes of the accessible one. I shall first show that the statement is correct.

We envisage the time-like world line of an observer on (H), reaching from the infinite past to the infinite future. We wish to know the range of his potential experience, by which we mean all those point-events that may ever enter his ken, that is to say, all those which he by his movements brings into his fore-cone. (An "event," of course, is a fixed point on (H) and does not move.) Once this has happened with a particular event P, he can never remove it from there by his future movements. (This is the same as our previous statement that he can never escape from the after-cone of P once he has entered it; the fore-cone of an observer all the time *only gains* points, the after-cone *only loses* points.) Hence his total potential experience is nothing else than the limit of his fore-cone in the infinite future. Now we have seen that in the remote outskirts of our model the (Euclidean) opening of the light-cone at, any point reduces ever more, tending to zero in the limit. All time-like world lines, hence also that of our observer, whatever it may have been in the earlier parts, approaches asymptotically to a meridian hyperbola – and the observer's velocity to that of tight; which seems strange but is due to the increasing expansion of "space" in our frame. Anyhow, the tangential plane (whose intersection with (H) is the light-cone!) approaches to a definite $45°$ – plane through the origin O of R_3, for example, in a particular case, to the plane $z = y$. Any such plane divides (H) into two equal halves; our observer's fore-cone approaches asymptotically to one of them, which therefore represents his total potential experience. (In our example it would be the half $y \geqslant z$.) It will be seen that it includes an almost complete knowledge of space in the remote past, but less and less about later spaces; of the bottle-neck parallel just one half becomes known in the limit; of the latest parallels ("spaces") only insignificant fractions. This is not amazing. Owing to the increasing rate of expansion, which asymptotically approaches that of light, that region of space from which light-signals can reach the observer *at all* contracts towards him. This *seems* to be in contrast with the known *equivalence* of all points on (H), hence also of all moments on the observer's world-line. But we must, remember, as was mentioned before, that the parallel circles, $z = $ constant, which play the part of space, or spatial cross-sections, are in our frame by no means equivalent to each other.[2]

---

[2]The objection might be raised: since all points on (H) are equivalent, it seems strange that the fore-cone of our observer in the distant future should include "very nearly half" of the surface, while, for example, in the remote past it includes much less – one is tempted to

Let me mention by the way that our present over-all result for any observer's potential experience affords an independent evidence of the inaccessibility of messages from antipodes to the same observer, which we proved directly above.

While any single observer can only explore one half of the "world", delimited as described above by some 45°-plane through the origin, he has still ample choice which half. Even if he starts from a given region in the remote

---

say only an "insignificant fraction". How strange that the fore-cone should be "growing", at all, since every point can by a Lorentz transformation be transformed into every other one, and thus, of course, its fore-cone into that of the other one! The answer is this. While the first two expressions in inverted commas may be dangerous and objectionable when applied to an *infinite* set, yet, it is certainly true that the observer's fore-cone grows, in the sense that at any later moment it, contains regions which it did not, contain at some earlier moment, *as well as* everything that it contained then ("earlier" and "later" may refer to the observer's eigen-time). This statement is invariant and does not militate against the equivalence. Compare it, with the simple case of a mass point moving uniformly on a straight line: no geometrical point in front of it can be indicated that will not be passed and left behind by the moving point at a time that, can bo indicated, even though any two situations during its motion are equivalent.

In careful, though somewhat circumstantial phrasing the theorem of the text runs thus: given the observer's world line, with any pair of antipodic events P, $\bar{P}$ (save a one-dimensional set to be indicated forthwith) a finite, time z_p can be associated, such that, for any $z > z_p$ either P or $\bar{P}$ is inside the observer's fore-cone. The excluded set are those situated on a certain couple of parallel generators g_1, g_2, that depend on the observer's world line. The association of z_p with P, $\bar{P}$ has not, the properties of uniform convergence; that is, z_p has no upper boundary. In all this statement, one may, if one prefers, replace the time z by the observer's eigen-time (but, of course, the value s_p, will not be tho same as z_p)

Let, me add a warning that is frequently disregarded. The freedom of using any frame of reference does not include the freedom of using more than one at a time, let, alone that of examining a course of events in a continuously changing frame.

60

past – provided you allow him to move nearly with the velocity of light – he may still choose to reach the distant, future at nearly any azimuth (meaning the angle *around,* the z-axis) within nearly 360°. This is easily seen by reversing the argument which has just shown us that, an observer in the distant future can be reached by messages from almost any azimuth in the remote past, with the only exception of a narrow vicinity of the azimuth antipodic to his own. Indeed, he is himself a message, and, being allowed to move with any velocity up to that of light, he might himself have started from nearly any azimuth in the remote past. Conversely if he starts from a *given* azimuth α in the remote past, he may direct himself to almost any azimuth in the distant future, with the only exclusion of a small neighbourhood of $\alpha \pm 180°$. This was our assertion.

It would not do therefore to delimit any particular half of the hyperboloid (H) as our model of the world, discarding the other half; nay, we definitely need the whole of it, if the model is to embrace the potential experience of any and every observer. By cutting out any portion, we should, erect unmotivated barriers to those time-like world lines (and null lines) that are about to penetrate into this portion; indeed, we should be at a loss to account, for what becomes of objects – particles, light-signals or observers – whose world lines they are.

On the other hand, it does seem rather odd that two or more observers, even such as "sat, on the same school bench" in the remote past, should in future, when they have "followed different paths in life", experience different worlds, so that eventually certain parts of the experienced world of one of them should remain *by principle* inaccessible to the other and vice versa. It is true that they can never become aware of this being so, except by theoretical considerations such as we are conducting just now, certainly not by comparing notes. For whenever they meet for this purpose, they have the same fore-cone and thus have at that moment reached the same potential experience. Still, we – or at any rate some of us – are used to regard the "real world around us" as a mental construct based precisely on the community of the experience of all normal, sane persons. From this point, of view one may find it distasteful to accept a world model according to which two observers who separate are likely to have the possible sharing of their experience stopped with regard to some parts of it that are interesting and relevant to them respectively. Indeed, this will happen sooner or later, unless they reach the distant future at precisely the same azimuth.

One way of avoiding this is to accept the elliptic interpretation of the world model (H); that is, by declaring antipodes to represent the same event. For then any half of (H), containing no antipodic pairs, amounts to the whole world. The total potential experience of any observer is complete and embraces therefore the same events for any two observers whatever their world lines be. But what price have we to pay for it?

I find it, convenient, though others may not need this help, to think now of an event as represented not just by the couple of antipodes, but by a thin, straight, rigid rod joining them. All these "rods" cross at the origin O. Envisage the succession of events that correspond to a mass point, or perhaps also to a light-signal. The "world line" of this object may be represented by the surface formed of the rods joining a couple of antipodic world lines (time-like or null lines; in the latter case the surface is plane). We feel tempted to allot *directions* to these two world lines, indicating the temporal succession by arrows. If so, the arrows must be opposite, one pointing to increasing z, one to decreasing z. But we must not do it. *For the ambiguity in this allotment is undecidable.* Just as we could

62

not decide on retaining only one "half" of the surface (H), casting the other one away, so we cannot decide on letting the arrows point "upward" in one half, "downward" in the antipodic half. For in whatever way we make the division there would be utter confusion for the world lines that cross it. So the distinction between past and future is lost, and, of course, also that between fore-cone and after-cone. In particular, we must not give any preference to "uni-directional" light-signals of one kind; we must, give advanced potentials the same standing as retarded potentials.

This is bewildering, but perhaps not, absolutely fatal to the elliptic model. It proves to be as completely *reversible* as all our fundamental theories are. It is capable of harbouring particles governed by reversible mechanical laws and reversible electromagnetic fields, including light-signals governed by reversible electrodynamics. In a model that includes time as a coordinate, the world line of a particle does not gradually come into being, it *is* the particle. To indicate its direction is not, necessary, if the laws are reversible, for it makes good sense either way. It is not necessary to know whether the motion is from A to B or from B to A; similarly with electromagnetic fields, including light-signals. Trouble arises when the moving something is a thermodynamical system, say an isolated system. In this case we are accustomed to decide, from the values of the entropy at A and at B, which is earlier. And, of course, the same trouble – and for the same reasons (if for no others) – arises in the case of an observer, who is (if nothing more) at any rate a thermodynamical system, carrying an innate arrow of time.

Still this may not be fatal to the elliptic interpretation. For it is well known that the irreversible laws of thermodynamics can only be based on the statistics of microscopically reversible systems on condition that the statistical theory be autonomous in defining the arrow of time. If any other laws of nature determine this arrow, the statistical theory collapses.[3] From this point of view a reversible world model is highly desirable.

I am afraid that I must leave it there for the moment, and turn to other aspects of de Sitter's world.

4. The static frame

The frames that we have used hitherto have the great advantage that the automorphisms of the model take the familiar form of Lorentz transformations of the embedding manifold. This simplicity is paid for by the need for employing a redundant coordinate, viz. three in the reduced model, five in the full model. Another inconvenience that we met is the inequivalence of the "spaces" z =constant. It is an essential inequivalence, since one of them ($z = 0$) is a geodesic, the others are not.

The simplest and not very profound way of scrapping the redundant coordinate is to introduce a couple of polar coordinates for x and y, or rather – since the spatial radius vector is not a constant – pseudospherical coordinates for x, y, z. If we put

$$x = R \cos \chi \cosh t, \quad y = R \sin \chi \cosh t, \quad z = R \sinh t,$$
$$(7)$$

this fulfills equation (H) and gives for the line element (6)

$$ds^2 = -R^2 \cosh^2 t d\chi^2 + R^2 dt^2. \qquad (8)$$

This is the same "contracting and expanding" world as before. The insignificant change is that we have introduced a new time t, which varies less rapidly than z. Notice that the whole surface (H) is embraced, if χ is given the range 2π and t all real numbers. The transformation (7)

---

[3]See E. Schrödinger, *Proc. R. Irish Acad.* **53**, 189, 1950.

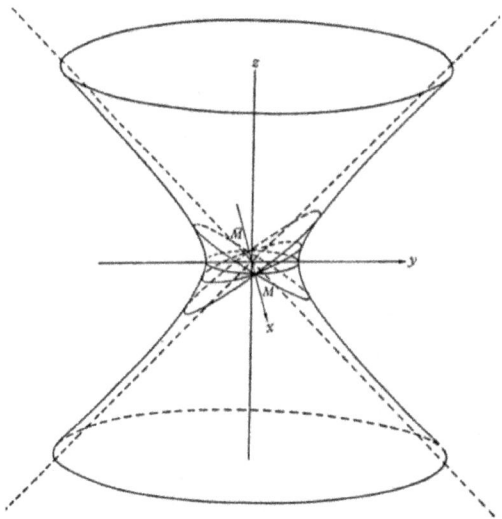

Fig. 2 The same as Fig. 1, but exhibiting a few of the space-like geodesics (ellipses) cut out by the bundle of planes passing through the x-axis. The two points M and $\overline{M}$ on this axis represent the so-called mass horizon of the static frame, the ellipses are the contemporary spaces in that frame.

is nowhere singular. The automorphisms are the changes (pseudo-rotations) from one pseudo-polar frame (χ, t) to another. It is easy to modify (7) for the full model (3), by two more polar angles θ, ϕ. We need not enlarge upon this. In (8)

$$d\chi^2 + \sin^2 \chi (d\theta^2 + \sin^2 \theta d\phi^2) \qquad (9)$$

would appear instead of just $d\chi^2$. It is the line element of a (three-dimensional) hypersphere, which replaces the *circle* of the reduced model. In what follows we keep to the reduced model.

In order to obtain a frame in which the spaces for constant time are all equivalent, we must follow another way. We remember that all the ellipses, cut out by planes through the origin with a slope of less than 45°, are equivalent space-like geodesics since each of them is carried by, suitable automorphism into the "bottle-neck," which is one of them. Select from them all those that pass through the same "horizontal" straight line, say the x-axis. We propose to make *them* the spaces for constant time. The new time must then be a function of the ratio z/y or, if you like, of the parameter of that Lorentz transformation in the zy-plane that would carry a particular ellipse into the "neck". A suitable choice seems to be

$$\tanh t = z/y. \qquad (10)$$

(This is precisely the 'velocity' of the said Lorentz transformation.) One suspects x or a function thereof to be a suitable space coordinate, because it is not affected by that Lorentz transformation. Let, us choose χ, where

$$\sin \chi = x/R. \qquad (11)$$

We comply with (10) and with the equation (H), if we supplement (11) thus:

$$x = R \sin \chi, \quad y = R \cos \chi \cosh t, \quad z = R \cos \chi \sinh t. \qquad (12)$$

We have reached an alternative set of pseudo-polar angles (χ, t) (H), very different from (7). The line element works out from (6) and (12)

$$ds^2 = -R^2 d\chi^2 + R^2 \cos^2 \chi dt^2. \qquad (13)$$

(For the full model the same modification as was indicated in (9) for (8) would occur in (13); but we keep to the reduced model, for simplicity and visualizability.) This is the familiar *static* form of the de Sitter metric, even more familiar in the coordinates x, t':

$$x = R \sin \chi, \quad t' = Rt,$$

which gives

$$ds^2 = -(1 - x^2/R^2)^{-1}dx^2 + (1 - x^2/R^2)dt'^2. \quad (14)$$

But we keep to (13). The static character is obtained at a very high price. The allotment (12) differs from (7) by two very incisive features. *First*, it does not avoid the singularity so well known from ordinary polar coordinates, unavoidable there. At, $x = \pm R$, $z = 0$ (points M and $\overline{M}$ in the figure), all spaces of simultaneity meet and the time t becomes entirely indeterminate, as can also be seen from (12) for $\chi = \pm 90°$. This is the notorious mass horizon, which here (and also in the full model, where it is a two-dimensional spherical surface) separates the domains $|\chi| > 90°$ and $|\chi| < 90°$ so efficiently, that they seem to have nothing to do with each other. *Secondly*, even with the fullest possibly useful range (i.e. χ from $-180°$ to $+180°$ and t all real values) (12) does not embrace the whole surface (H), but only that part in which $|y| \geqslant |z|$, as can be seen from (10). The limits are the two "light-cones" (four null lines) through the points $x = \pm R, z = 0$.

These two circumstances together, the singularity and the limitation, both artificially produced by the choice of frame, have strange consequences. Let us envisage on the hyperboloid (H) the saddle-shaped section to which the representation (12) is really confined, given by $|x| \leqslant R, y \geqslant 0$, or $|\chi| \leqslant 90°$, t real. (It would make no difference if we included the antipodic saddle $|\chi| > 180$, but it is sufficient to speak of one of them.) Remember that the time-like geodesics – which represent free test particles – are traced on (H) by planes through the origin with a slope steeper than 45°, and have themselves, of course, an ascent steeper than 45°. It will be realized that most of them have only a finite section inside our "saddle" region. A notable exception is the hyperbola branch $x = 0$, which in the static frame represents a particle permanently at rest at the spatial origin ($\chi = O$) of this frame. This orbit is notable also because it is the "Greenwich" of the static frame; its eigen-time, from (13), is Rt and thus determines the world time t. No other free-particle orbit passes its whole course within our region, though there are still two groups, asymptotic to one or other branch of the Greenwich hyperbola, which either have been inside from $t = -\infty$, leaving the region later or, conversely, enter it and remain inside for ever. The latter represent particles that move towards the static origin and come to rest there asymptotically, the former represent precisely the reverse motion.

Even in these exceptional cases the entrance or the escape, respectively, occur at a finite *eigen-time* of the particle in question. In the general case, when only a finite portion of the orbit lies inside our region, the particle passes only a finite stretch of its *eigen-time* inside. All this is obvious in the frame used on the hyperboloid (H) – and the eigen-time is an invariant. We may add that light-signals (generators) also enter and leave the saddle-shaped region and have a continuation both ways outside.

How do these happenings present, themselves in the *static frame?* The description usually given is this.[4] A particle, or a light-signal, may be found approaching the origin and may continue to do so for some time. Eventually (save for the very exceptional case mentioned above) it will turn away from it, never to return. From then onwards it continues to approach the "mass horizon", but reaches it only asymptotically for $t \to \infty$.[5]

---

[4]See, for example, R. C. Tolman, *Relativity, Thermodynamics and, Cosmology* (Oxford University Press, 1934), pp. 349ff.

[5]EDITOR'S NOTE: Unfortunately, expressions like this – "reaches it only asymptotically for $t \to \infty$" – are a source of confusion in spacetime physics. As explained in [1] the very concept of a black hole is a result of using double standards in the interpretation of

The most bewildering trait of these statements, when held against, the (H) model, is this. The mass horizon consists of the two coordinate singularities at the points $M, \overline{M}(x = \pm R, z = 0)$They are joined by a *space-like* geodesic to every point within our saddle-shaped region. How should anything ever even get near one of them? Besides we know, and see by direct inspection, that most time-like (and null) geodesics enter and leave our region at quite different points, and without being in the least hampered or impeded by the old mass horizon from crossing the boundary.

The solution of this paradox is that the boundary of our region is formed by parts of the *null* lines through either of the two singularities at $M, \overline{M}$. In approaching such a null line you find that your invariant distance from that singular point, while remaining purely spatial, approaches to zero. (It is a quaint feature of the relativity metric that, on the extremely oblong ellipses which you reach, the long, nearly straight, parts of the arc, being nearly null lines, contribute very little to the total circumference, which accrues mainly from the short bits at the turning points near the apices of the major axis.)

The alleged inability of the movable object, whether a mass point or a light-signal, in the static frame to reach, let alone to cross, the border is merely a matter of the "Greenwich" time definition. It is on the same footing as

---

Schwarzschild's solution:

- light needs infinite time to leave a black hole – "infinite time" in this case is interpreted as never.

- black holes need infinite time to form for distant observers (like all of us) – "infinite time" in this case is *not* interpreted as never and black holes are assumed to exist for us (as distant observers).

1. V. Petkov, *Seven Fundamental Concepts in Spacetime*, SpringerBriefs in Physics (Springer, Heidelberg 2021), Ch. 7.

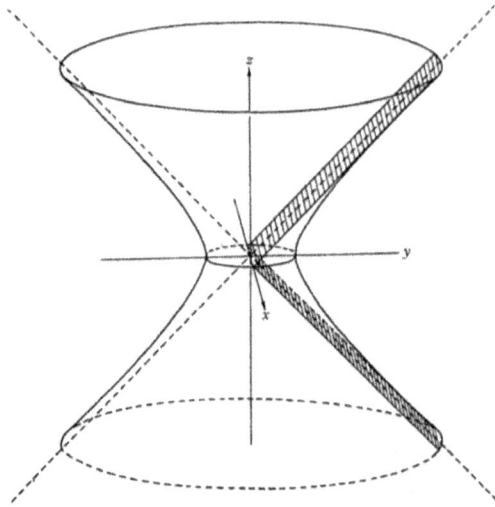

Fig. 3 The same as Figs. 1 and 2, but exhibiting the "saddle-shaped" region to which the whole world is restricted in the static frame. It is *that* part of the surface of the hyperboloid that, lies on the right of the two hatched plane strips. Its boundaries on the surface are four (halves of) null geodesics issuing from the two points $M, \overline{M}$ (the mass horizon). These null geodesics are cut out by the planes $x = \pm R$. Please realize that there exist many time-like and null geodesics (none of them drawn in the figure in order not to confuse it) that enter and leave this world, spending not more than a finite interval of eigen-time inside it.

the alleged inability of Achilles to outrun the tortoise in Zeno's paradox. The time t is defined so as to become either ∞ or $-\infty$ on the borders, hence they cannot, be crossed at any finite time.

Some authors hold, or at least favour, the view that the static frame is that of an "observer" permanently at rest at, the spatial origin (the hyperbola branch $\chi = 0$, or $x = 0$, $y > 0$. Their motive is obviously that this is the only time-like geodesic which remains permanently, "from dawn to twilight", inside the picture and that it seems

70

distasteful to find oneself drop out of one's world frame. But there is no earthly reason for compelling anybody to change the frame of reference he uses in his computations whenever he takes a walk.[6] Even in the very much simpler case of ordinary civil time we do not keep changing our watches when we travel around the British Isles or France.

But even if our "observer" does remain at the spatial origin, he is bound to find out theoretically – provided he has established his static world model correctly and uses it correctly – that particles, light-signals, and observers other than himself, usually travel from one part of the mass horizon to another within a finite stretch of their *eigen-times*, and cross it, without any effort of their own, just by ceding to the inertial motion. He finds this out in the case of objects whose orbits he has observed for a while through his telescopes with sufficient, accuracy for computing, from the known laws of inertial motion, their whole courses, and also the finite period of *eigen-time* that such an object has spent in his world and will spend before it *reaches* the mass horizon again. He is bound to wonder where that object has come from and where it is going to.

And more than that. He may launch a rocket and compute from the initial velocity that he imparts to it the *finite* eigen-time of the rocket when it will reach the horizon. He has put into the rocket a thermodynamic system in non-equilibrium and a quantity of uranium. The *return* to equilibrium of the former and the *decay* of the latter is governed by the eigen-time. Hence the return may not,

---

[6]Let, me on this occasion denounce the abuse which has crept in from popular *exposés*, viz. to connect any particular frame of reference, e.g. in special relativity, with the behaviour (motion) of him who uses it. The physicist's whereabouts are his private affair. It is the very gist of relativity that anybody may use any frame. Indeed, we study, for example, particle collisions alternately in the laboratory frame and in the centre-of-mass frame without having to board a supersonic aeroplane in the latter case.

and the decay certainly will not, be completed when the horizon is reached. But this is at the *world*-time $t \to \infty$. Thus in the static frame, because the eigen-time of the central geodesic is adopted as world time,[7] those physical processes appear to be deferred indefinitely. (The current description is that they are slowed down indefinitely as the systems approach the horizon.)

The converse occurrence, a rocket *reaching* the central observer, presents less, or at any rate no new, interest. But here the case of *light-signals* deserves to be examined, since they may have been emitted by an event which is situated outside the saddle-shaped region on the hyperboloid (H), and would in this case vouchsafe to the central observer information about, an event to which he must not allot any real value of t. Eddington's[8] sharp intellect noticed this a long time ago: "he might even *see* them [i.e. such

---

[7]EDITOR'S NOTE: Here the meaning of the concept "world time" is clear, but similar concepts (e.g., a preferred or rest frame, which imply preferred time) often cause confusion in cosmology. Here is the essence of the confusion-free explanation by B. Schutz in his book *Gravity from the ground up*, p. 357 (and the whole chapter there):

> It is interesting to reflect that relativity began with the assumption that we could never measure our own velocity, not in an absolute sense. We discussed this in Chapter 15. Has relativity now led to the opposite, a measurement of our velocity in the Universe? There is no real contradiction here, but the issue is intriguing nevertheless. The velocity we measure from the cosmic microwave background is, like every other velocity in relativity, a velocity relative to something. In this case, it is relative to the mean rest frame of matter at the time the universe became transparent to photons. We only determined this velocity by making a measurement on something outside of ourselves: the radiation itself. Velocities are still relative, even in cosmology.

[8]*The Mathematical Theory of Relativity*, 2nd ed. (Cambridge University Press, 1930), §70, p. 166. Eddington calls the static frame a "make-shift contrivance" that "leads us to the admission of

events] happening through a powerful telescope."

This in itself is correct. But whether it would really cause embarrassment to our observer from his point of view must be examined. Unfortunately, our *reduced* model is insufficient for this purpose. To begin with, a single event emits only one light-ray to a single observer – irrespective of whether space has one or three dimensions. From this one ray the observer cannot locate the source; it may be anywhere along the ray. At least two observers at some distance are required to determine the parallax. But clearly in one-dimensional space this is of no avail. There is no parallax, the signal can only arrive from two directions, either from the right or from the left, i.e. either from the direction of M or of $\overline{M}$.

So we must now go back to our model (3) embedded in a Minkowski R_5 with the metric

$$ds^2 = -dx^2 - dy^2 - du^2 - dv^2 + dz^2, \qquad (15)$$

and to a corresponding static model.

5. The determination of parallaxes

Let, me sum up: we have seen tha when we transform our reduced model from the xyz-frame to the static frame, the latter embraces only a fractional region of the whole hyperboloid, bordered by the two points M, $\overline{M}$ and four null lines issuing from them. It is reduced to this region because a time reckoning has been introduced such that t becomes $+\infty$ or $-\infty$ on these null lines and indefinite at the points M, $\overline{M}$, the mass horizon. This leads, as we have seen, to strange happenings, because objects and light-rays enter and leave this region, and most objects pass only a finite stretch of eigen-time within. We have ascertained

---

extra-temporal events as affecting even our own experience".

73

that *any* object (or light-signal) when entering or leaving is projected to the mass horizon in this frame. But in the reduced frame, space being one-dimensional, there are no parallaxes and therefore an observer would not be able to locate the place of origin of such an external light-signal. We must therefore now go back to the full-fledged model with x, u, v for space, in lieu of just x.

It is easy to see – and I shall show that afterwards in a few lines – that x, y, v can be used as space coordinates in the static frame. But first, in order not, to confuse the frames, we shall ascertain in the five-dimensional *xuvyz*-fame to which points on their mass horizon "static observers" near the static origin will project light-signals coming to them from the same external event. (Afterwards we shall ponder about how they will interpret them.)

For the spatial origin (or world line of the central observer) in the static model we choose

$$x = u = v = 0, \quad y > 0, \tag{16}$$

actually the same as in the model reduced by (5). The mass horizon corresponding to this choice is traced on the hyperhyperboloid (3)

$$x^2 + u^2 + v^2 + y^2 - z^2 = R^2 \tag{3}$$

by

$$y = 0, \quad z = 0, \tag{17}$$

which in the reduced model determined the reduced mass horizon $M, \overline{M}$ ($x = \pm R$, $z = 0$); the full one, determined by (3) and (17), can be visualized as a two-dimensional sphere

$$x^2 + u^2 + v^2 = R^2 \tag{18}$$

in the sub-R_3 spanned by x, u, v. What in the reduced model we called the saddle-shaped region, which was the *world,* of the static model, is now the four-dimensional region

$$x^2 + u^2 + v^2 \leqslant R^2 \qquad y \geqslant 0 \qquad (19)$$

on the hypersurface (3).

The boundary that one has to pierce to get inside this region from without is of course not just the mass horizon (17) – a four-dimensional region wants a three-dimensional boundary – the boundary is the R_3 traced on (3) by (18). Combining these two equations you see that on the boundary $|y| = |z|$. The boundary consists of certain null lines coming from the points on the mass horizon. Indeed, if the coordinates of a point, on the latter are indicated by the subscript M, the equation of its light-cone,[9] written in the running coordinates with a dash, reads, from (17),

$$-(x'-x_M)^2-(u'-u_M)^2-(v'-v_M)^2-y'^2+z'^2=0 \quad (20)$$

It will be seen that, the couple of generators of this light-cone

$$y' = \pm z', \quad x' = x_M, \quad u' = u_M, \quad v' = v_M \qquad (21)$$

lie on the hyper-surface (3) *and,* on the boundary, *and no other point does,* since the canceling of the terms with y' and z' in (20) necessitates the last three equations in (21). We make a special note of the fact that every point on the boundary is connected by a null geodesic with one, and only one, point on the mass horizon, viz. with the one which has the same x, u, v but, from (I7), vanishing y, z.

Now we want to choose an event $E(x_E, y_E, \text{etc.})$ *out-side* that region, such that a light-ray from E reaches the hyperbola branch (16). We take it so that at E the first inequality (19) is violated (but not the second) and take

---

[9]The full one in five dimensions!

$z_E < 0$. For simplicity we choose E on the fore-cone of the apex of (16) which apex is

$$x = u = v = z = 0, \quad y = R. \tag{22}$$

Its fore-cone,[10] written in running coordinates x'', etc., is

$$-x''^2 - u''^2 - v''^2 - (y'' - R)^2 + z''^2 = 0, \quad z'' \leqslant 0. \tag{23}$$

Since the x'', etc., have to satisfy (3) this is reduced to

$$y'' = R, \quad z'' \leqslant 0, \tag{24}$$

while x'', u'', v'' take all the values compatible with

$$x''^2 + u''^2 + v''^2 = z''^2. \tag{25}$$

We choose our event E as follows,

$$x_E^2 + u_E^2 + v_E^2 = z_E^2. \quad y_E = R, \quad z_E < -R, \tag{26}$$

the latter in order to violate the first inequality (19). The light-ray going from E to the apex (22) is obviously

$$\begin{aligned} x'' = \lambda x_E, \quad u'' = \lambda u_E, \quad v'' = \lambda v_E, \\ y'' = R, \quad z'' = \lambda z_E, \end{aligned} \tag{27}$$

where λ is a parameter going from $\lambda = 1$ (at E) to $\lambda = 0$ (at the apex). This light-ray pierces the boundary at a point where $|z''| = |y''|$, thus at

$$z'' = \lambda z_E = -R, \quad \lambda = -R/z_E = \lambda_0 \quad \text{(say)} \tag{28}$$

λ_0 being a positive number between 1 and 0, on account of the inequality (26).

---

[10] The full one in five dimensions!

This piercing point is, as we have seen, joined by a null geodesic with one and only one point on the mass horizon, viz. to the point

$$x_M = \lambda_0 x_E, \quad u_M = \lambda_0 u_E,$$
$$v_M = \lambda_0 v_E, \quad y_M = z_M = 0. \tag{29}$$

And this is, in the central observer's static frame, the point to which he projects the event, E, or where he "sees" it.

Now we want to consider the result of a parallax measurement by him and a neighbouring observer. We pick out, near the previous piercing point, a point on the boundary with coordinates

$$x = \lambda_0 x_E + \Delta x$$
$$u = \lambda_0 u_E + \Delta u$$
$$v = \lambda_0 v_E + \Delta v \tag{30}$$
$$y = -z = R + \Delta y,$$

which is certainly possible though the first three increments are subject to (18). This secures also that, the new piercing point (30) satisfies (3). The increment Δy shall be chosen so that the straight line joining E with (30) is a light-ray, which can easily be proved to be possible. (We might have started from considering a neighbouring light-ray from E and defined (30) as the point where it pierces the boundary.) The coordinates of this light-ray, say x''', etc., have the following parameter representation, since the ray connects the points (26) and (30):

$$x''' = x_E + \mu[(\lambda_0 - 1)x_E + \Delta x],$$
$$u''' = u_E + \mu[(\lambda_0 - 1)u_E + \Delta u],$$
$$v''' = v_E + \mu[(\lambda_0 - 1)v_E + \Delta v], \tag{31}$$
$$y''' = R + \mu \Delta y,$$
$$z''' = z_E + \mu(-R - z_E - \Delta y).$$

Indeed, for $\mu = 0$ they coincide with those of E, for $\mu = 1$ with (30). For the following μ-value

$$\mu = \frac{1}{1 - \lambda_0} = \frac{z_E}{z_E + R}$$

(see (28)), which is greater than 1 (see the inequality (26)), they are

$$x''' = \frac{\Delta x}{1 - \lambda_0}, \quad u''' = \frac{\Delta u}{1 - \lambda_0}, \quad v''' = \frac{\Delta v}{1 - \lambda_0}$$

$$y''' = R + \frac{\Delta y}{1 - \lambda_0}, \quad z''' = -\frac{\Delta y}{1 - \lambda_0}. \tag{32}$$

This is close to the apex (22). We assume an auxiliary observer at (32), posted there for the purpose of measuring the parallax. According to what has been said, he "projects" the ray to that point of the mass horizon which has the same x, u, v, as the piercing point (30). Notice that it is the *same* mass horizon! For the two observers needs must use the same static frame if their comparing of notes is to be meaningful. From (30) the auxiliary observer finds his point of projection on the horizon displaced by $\Delta x, \Delta u, \Delta v$, with respect to the finding of the main observer, while he knows his own observatory to be (from (32)) displaced in the same spatial direction by proportional amounts, but larger by the factor $1/(1 - \lambda_0)$, with respect to the main observer's observatory. Since the two spatial displacements are parallel, the latter displacement, is in the static frame orthogonal to the line of vision, which facilitates the computation of the parallax. The displacement on the horizon being in the same (not the opposite) direction and smaller by the factor $1 - \lambda_0$, the two observers on comparing notes will locate the event E behind the horizon at a distance (from the horizon) that is easily computed to be $(1 - \lambda_0)R/\lambda_0$. Since according to their time reckoning the horizon cannot be reached, let alone traversed, in a finite world time, they would find it impossible to allot any value of time to the event E, as Eddington has pointed out.

Now the preceding considerations are flawless, as far as the computation of the displacements, Δx etc. and $\frac{1}{1-\lambda_0}\Delta x$ etc. is concerned; but in considering the parallax that would arise from it, and the conclusion the observers would draw from it I have been negligent. First, I have used a fact (which is derived, for instance, by R. C. Tolman, *op. cit.* pp. 349ff.), viz. that light-rays are straight in the static frame, precisely if x, u, v are used. This is a little strange, because static space is not flat. So we must look into that. The full transformation corresponding to the reduced one given in (12) reads:

$$x = R\sin\theta\cos\phi\sin\chi, \quad u = R\sin\theta\sin\phi\sin\chi,$$
$$v = R\cos\theta\sin\chi, \quad y = R\cos\chi\cosh t, \quad z = R\cos\chi\sinh t.$$
$$(12a)$$

This satisfies (3), the hyper-hyperboloid. The line element becomes (cf. (13))

$$ds^2 = -R^2[d\chi^2 + \sin^2\chi(d\theta^2 + \sin^2\theta d\phi^2)] + R^2\cos^2\chi dt^2.$$
$$(13a)$$

Space is obviously hyperspherical, radius R. But, one may also use x, u, v as spatial coordinates, putting

$$r = R\sin\chi, \quad x = r\sin\theta\cos\phi, \text{ etc.}, \quad r^2 = x^2 + u^2 + v^2.$$

Then (13a) becomes

$$ds^2 = -(1 - r^2/R^2)^{-1}dr^2 - r^2(d\theta^2 + \sin^2\theta d\phi^2)$$
$$+ (1 - r^2/R^2)R^2 dt^2.$$
$$(14a)$$

Thus x, u, v and r, θ, ϕ, are ordinary Cartesian and polar coordinates respectively, with the familiar relations between them, but the space geometry is Riemannian, radial line elements are 'longer' than in flat space, e.g. the invariant distance of the horizon ($r = R$) is not R, but

$$\int_0^R \frac{dr}{\cos \chi} = \int_0^{(1/2)\pi} \frac{R \cos \chi d\chi}{\cos \chi} = \frac{\pi}{2} R.$$

Now what is the shape of a null geodesic? First, on (3) in R_5 they are traced by certain *planes* through the origin O of R_5. Thus they are plane curves. Moreover, they must, have a constant inclination to the z-axis. *Hence they must be straight lines in R_5.* This implies four independent linear equations in the five variables x, u, v, y, z (not all four homogeneous, by the way, since the lines do not pass through the five-dimensional origin O, which is outside the hyper-hyperboloid). By eliminating y and z we get, two linear equations in x, u, v. Thus the result that *the light-rays are straight lines in xuv-space.* We know this space to have hyperspherical metric. It is not difficult to show that they are *not* spatial geodesics of this metric – that is, because the time flow[11] is not, uniform.[12] They are circles, but *not* great circles. *So we have projected, correctly.* Our

---

[11]EDITOR'S NOTE: Using the concept "time flow" in physics papers and books can give rise to misconceptions shared even by physicists. The idea that time flows comes from the way we perceive the world – we are aware of ourselves and the world only at one single moment of time – the present moment (the moment 'now') – and implicitly assume that only that moment exists and constantly changes, i.e., that time flows. However, there does not exist even a hint of experimental support for such an assumption. On the contrary, in the beginning of the twentieth century it was realized that the experimental evidence at that time unequivocally demonstrated that all moments of time are real (at therefore there is no flow of time) – in 1908 in his world-view-changing lecture "Space and Time" [1] Hermann Minkowski reported the profound physical message hidden in the failed experiments to detect uniform motion with respect to space – the reason those experiments failed was that there is no such thing as space in the physical world; Minkowski demonstrated that reality is a four-dimensional world with time as the fourth dimension (which means that all moments of time have equal existence exactly like the space points forming the three spacial dimensions). Minkowski stressed that the new spacetime view of the world is firmly supported by experimental physics – indeed, all negative experimental results mentioned above (and captured in Galileo's principle of relativity) and length contraction *would be impossible if spacetime were not real* (more precisely, if the concept of spacetime did not represent a real four-dimensional world).

Most probably Minkowski would point out that one cannot talk about time flow in spacetime because that constitutes a clear contradiction in terms – time flow implies that only the moment 'now' is real, whereas all moments of time are real in spacetime (more precisely, all events of spacetime are real).

1. H. Minkowski, Space and Time, in: *Spacetime: Minkowski's Papers on Spacetime Physics* (Minkowski Institute Press, Montreal 2020).

[12]In other words, because tho velocity of light differs from place to place and Fermat's principle of shortest *time* (not shortest path) holds.

Please observe that there is no contradiction between the two statements about the light-rays being *straight lines* in x, u, v , but *circles* when this space is contemplated as (half of) a hypersphere.

observers will measure a *positive* parallax. That we must regard as a *fact* (in such a world).

But there is a second question: have we made them interpret their parallax intelligently? Knowing that light follows straight lines (in x, u, v), it would not, be unnatural to surmise that this holds also beyond the horizon, and that was the tacit assumption I made above in 'locating' the event E for them. But there is an impasse in this. Along a ray from the spatial origin outward x, u, v and r grow proportionally. At the horizon $r = R$. If you continue on the ray beyond, then $r^2 = x^2 + u^2 + v^2 > R^2$. Such points are not to be found, if you want space to continue hyperspherically also beyond the half-sphere that is inside the horizon. And this is the most natural assumption – but in obvious contradiction to the first natural assumption. I do not see how the problem of location can be intelligently solved.

But we note the result, that they do get a reasonable positive parallax, because we shall need it later: the actual setting of their telescopes cannot, depend on the frame we or they use.

6. The Lemaître-Robertson frame

This frame is interesting because it presents space as infinite and, *qua* space, flat, showing that the question asked so often, whether space is really curved and finite, cannot be answered, because the answer depends on the frame. We turn at first to our reduced model. The salient point is that, we propose to regard as 'contemporary spaces' the set of parabolae traced on (H)

$$x^2 + y^2 - z^2 = R^2 \tag{H}$$

by set of parallel 45°-planes, and that we choose them parallel to the x-axis, viz. the family

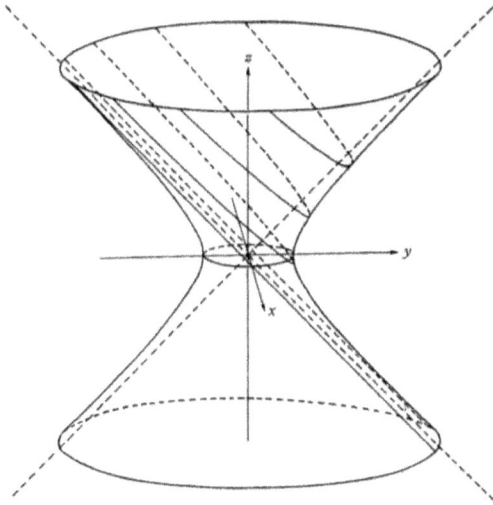

Fig. 4 The same as Figs. 1-3, but exhibiting the family of parabolae cut out by a bundle of 45°-planes, all parallel to the one through the x-axis. The latter dissects the hyperboloid by two null geodesics, into two equal halves, the upper half representing the whole world in the Lemaître-Robertson frame, in which the parabolae are the contemporary spaces. Their orthogonal trajectories (not drawn) are the time lines. They are cut out by the bundle of planes that pass through tho asymptote which has a slope from the left above to the right below (equations $x = 0$, $y + z = 0$).

$$y + z = \text{constant.} \qquad (33)$$

This means that we introduce a time variable

$$\bar{t} = f(y + z). \qquad (34)$$

This seems at first rather daring; but, indeed, the parabolae are all space-like (having a slope less than 45° everywhere) with one exception, viz. when the constant in (33) is zero, which gives a couple of parallel generators (null lines), viz. $x = \pm R$. This limiting case will be removed

83

later by a suitable choice of the function f. As in the previous model we are left with x for labeling the points in any one of these spaces, which are one-dimensional in our reduced model. (In the full model x, u, v will be used.) Of course we want the line element to be orthogonal; that is to say, if we take the space coordinate

$$\bar{r} = g(x, y, z), \qquad (35)$$

we wish the lines, $\bar{t} = $ constant and $\bar{r} = $ constant, to be (Minkowskian) orthogonal. Naturally that leaves no significant choice for $\bar{r}$, because there can be but one set of orthogonal trajectories to our parabolae. Let us determine them. The directions of $\bar{t} = $ constant on (H) are given by

$$dy + dz = 0,$$
$$x\,dx + y\,dy - z\,dz = 0. \qquad (36)$$

Thus
$$dx : dy : dz = -(y + z) : x : (-x). \qquad (37)$$

Any direction $\delta x, \delta y, \delta z$, that is Minkowski orthogonal to this one must satisfy

$$- (y + z)\delta x + x\delta y + x\delta z = 0, \qquad (38)$$

$$\frac{\delta(y + z)}{y + z} - \frac{\delta x}{x} = 0, \qquad (39)$$

$$\frac{x}{y + z} = \alpha (= \text{constant}). \qquad (40)$$

So we have to take[13]

$$\bar{r} = g\left(\frac{x}{y + z}\right). \qquad (41)$$

---

[13]By a slightly more involved procedure g can be determined directly and turns out to be an arbitrary function of the *two* arguments $x/(y + z)$ and $x^2 + y^2 - z^2$. But the second is constant on (H).

The planes (40), which trace the coordinate lines $\bar{r} =$ constant on the hyperboloid (H) pass through the spatial origin O. They are the bundle through the asymptote $x = 0, y + z = 0$. Since the planes pass through O, the coordinate lines are geodesics and, of course, time-like, as can be seen directly from (40); for the cosine of the angle between the z-axis and the normal to this plane is

$$\frac{|\alpha|}{\sqrt{(1 + 2\alpha^2)}} < \frac{1}{\sqrt{2}} \tag{42}$$

for any not infinite α. That the coordinate lines are geodesics is an attractive feature. It means that a free particle may remain 'at rest,' ($\bar{r} =$ constant) in this frame (which it could not do in the static frame, except at the spatial origin).

The simplest specification of (41) and (34), which we shall adopt, is

$$\bar{r} = \frac{Rx}{y + z},$$
$$\bar{t} = \log \frac{y + z}{R} \tag{43}$$

The choice of the logarithm removes the uncanny degenerate parabola $(y + z = 0)$ to the negative infinity of time. This can hardly be avoided, but is paid for by the loss of half of the hyperboloid (H), viz. where $y + z < 0$. We shall come back to this deficiency later. We note that $\bar{r}$ (in spite of the letter chosen) can have either sign, namely, that of x; its range is over all real numbers. As we approach $\bar{t} \to \infty$, only a 'small' neighbourhood of the space parabola's apex has finite values of $\bar{r}$, everywhere else $\bar{r}$ becomes very large. At $y + z = 0$ the frame is singular, the labels (43) are useless.

Inverting (43) one gets first

$$y + z = Re^{\bar{t}}$$

$$y - z = Re^{-\bar{t}} - \frac{\bar{r}^2}{R}e^{\bar{t}}, \qquad (44)$$

$$x = \bar{r}e^{\bar{t}}.$$

The second is obtained from (H), which gives

$$y - z = \frac{R^2 - x^2}{y + z}, \qquad (45)$$

by using here the first and last of (44).

From (44) the line element

$$ds^2 = -dx^2 - dy^2 + dz^2 = -e^{2\bar{t}}d\bar{r}^2 + R^2 d\bar{t}^2. \qquad (46)$$

'Space' is expanding, inasmuch as free test particles with constant values of $\bar{r}$ (which we have seen to be admissible) increase their mutual distances exponentially as time proceeds; at any fixed time, space is infinite and flat, though flatness is not very,significant in the reduced model. (But it is flat also in the full model.[14] The $\bar{r}$ is a uniform measure of time, the same everywhere and always. But, of course, the velocity of light,

$$\frac{d\bar{r}}{d\bar{t}} = Re^{-\bar{t}}, \qquad (47)$$

decreases rapidly as time proceeds, suggesting perhaps that it, is 'actually' constant and light is only delayed by the expansion of space.

We have seen that a test, particle can let itself be swept along with the expansion. Can it also resist it? No, not for

---

[14]Notice that no spatial origin is distinguished; you may shift $\bar{r}$ by a constant. But this transformation, here very simple, does not seem to correspond to a simple automorphism of (H).

very long. This is an amazing aspect of this frame. It can be inferred directly from the expression for $\bar{r}$ in (43), at least in our reduced model, for which we have seen much earlier that any time-like geodesic eventually approaches asymptotically to a meridian hyperbola branch, nay to its asymptote (though this asymptote is outside (H)). It is intuitively clear that then the ratios $x : y : z$ approach to constants and so must $\bar{r}$, which depends only on these ratios. The logarithmic time scale makes this approach appear very rapid, viz. exponential. Thus a particle with peculiar motion (change of $\bar{r}$) stops dead very soon, to be swept along with the expansion of space. *A light-signal makes no exception,* since the above consideration holds for it. A coordinate value $\bar{r}$ can be indicated which it will never surpass and only reach asymptotically, in the same way as any material particle. Considering (47) this is not astonishing. We shall give the analytical details for the full model, which we must, now consult, for two reasons. First, it is by no means clear, whether this stopping dead holds also for the tangential motion or perhaps only for the radial one. Secondly, we wish to know the shape of the orbits of free test particles. It will turn out that they are straight lines and really stop dead entirely.

For the full model the Lemaître transformation corresponding to (43) reads

$$\bar{x} = \frac{Rx}{y+z}, \quad \bar{u} = \frac{Ru}{y+z}, \quad \bar{v} = \frac{Rv}{y+z}, \quad \bar{t} = \log\frac{y+z}{R}. \quad (48)$$

We shall now use $\bar{r}$ for the spatial radius vector, thus

$$\bar{r}^2 = \bar{x}^2 + \bar{u}^2 + \bar{v}^2, \quad \bar{r} \geqslant 0. \quad (49)$$

The inverse equations read

$$y + z = Re^{\bar{t}},$$

$$y - z = Re^{-\bar{t}} - \frac{\bar{r}^2}{R}e^{\bar{t}}, \tag{50}$$

$$x = \bar{x}e^{\bar{t}}, \quad u = \bar{u}e^{\bar{t}}, \quad v = \bar{v}e^{\bar{t}}.$$

In the second we have used (3)

$$y - z = \frac{R^2 - x^2 - u^2 - v^2}{y + z}. \tag{51}$$

One easily computes

$$ds^2 = -dx^2 - du^2 - dv^2 - dy^2 + dz^2$$
$$= -e^{2\bar{t}}(d\bar{x}^2 + d\bar{u}^2 + d\bar{v}^2) + R^2 d\bar{t}^2, \tag{52}$$

confirming that the contemporary spaces $(\bar{x}, \bar{u}, \bar{v})$ are flat and of course, infinite. The whole Euclidean six-parameter group of movements in these three variables leaves (52) unchanged, no spatial point and no direction is distinguished. *Any* geodesic is found as the intersection of the hyper-hyperboloid (3) with a plane passing through the origin O of the five-dimensional embedding space x, u, v, y, z, i.e. by positing three independent homogeneous linear equations between these five variables. From them we express x, u, v by $y + z$ and $y - z$, then multiply by $R/(y+z)$, and get from the first three of (48)

$$\bar{x} = A + B\frac{y - z}{y + z}, \tag{53}$$

and two similar ones for $\bar{u}, \bar{v}$, with constants A', B' and A'', B'', say. Using the first two of (50)

$$\bar{x} = A + B\left(e^{-2\bar{t}} - \frac{\bar{r}^2}{R^2}\right), \tag{54}$$

88

and two similar. Eliminating the time, we obtain in any case at least two *linear* equations. In general

$$\frac{\bar{x} - A}{B} = \frac{\bar{u} - A'}{B'} = \frac{\bar{v} - A''}{B''}. \tag{55}$$

Of these two equations one or both may fail because one or two, respectively, of the B's may be zero. But in these cases one or two, respectively, of (54) are already time-free. All three of them are, if all three B's are zero; the orbit degenerates into a point, indicating a particle 'at rest', i.e. without peculiar motion. (That this is never a space-like geodesic is obvious from (52), because in this case

$$ds^2 = R^2 d\bar{t}^2 \geqslant 0.$$

The result that a geodesic is *straight* in $\bar{x}, \bar{u}, \bar{v}$ holds for all of them, but is mainly of interest for the *orbits* of particles and light-signals. We wish to prove that on a non-space-like geodesic $d\bar{t}$ can vanish nowhere (hence $\bar{t}$ ranges from $-\infty$ to $+\infty$). Indeed, from (52), if $d\bar{t}$ vanished anywhere for a, progress along a non-space-like geodesic, $d\bar{x}, d\bar{u}, d\bar{v}$ would have to vanish as well, lest ds^2 be negative (space-like). This, from (48), would entail

$$d(y + z) = 0, \quad dx = du = dv = 0.$$

Taking account of this and differentiating (3)

$$(y + z)d(y - z) = 0$$

If we except the region $Y + z = 0$, which is singular in the Lemaître frame, this entails

$$dy = dz = 0.$$

But with all *five* differentials vanishing we are not progressing at all along our geodesic. Hence, $\bar{t}$, since its differential cannot vanish, must range from $-\infty$ to $+\infty$ on a non-spatial geodesic, i.e. on any *orbit*. (In the reduced model this is intuitively obvious.)

It is now seen from (54) and the two similar equations that in the limit $\bar{t} \to \infty$ the peculiar motion stops dead rapidly in all three directions, the limiting values of $\bar{x}, \bar{u}, \bar{v}$, being one of the two sets of roots of the equation

$$\bar{x} = A - B\frac{\bar{r}^2}{R^2} \tag{56}$$

and the two similar ones. They are in general quadratic on account of (49), but, we have seen that they imply two linear ones – see (55) and the remarks following it. Though spatial geodesics are of less interest, it is well to point out that this proof does not hold (and the fact of the stopping dead is not true) for a spatial geodesic; and this for the simple reason that on it $\bar{t}$ does not go to $+\infty$ but has a maximum value. We have seen this intuitively in the reduced model. To prove it for the full model, we observe that the (Euclidean) cosine of the angle between the z-axis and the radius vector from O to any point x, u, y, z is

$$\frac{1.z}{\sqrt{(x^2 + u^2 + v^2 + z^2)}} = \frac{z}{\sqrt{(R^2 + 2z^2)}}.$$

It approaches to $1/\sqrt{2}$, from below, hence the said angle approaches to $45°$ from above, as z increases, and therefore z must have an upper boundary, if the point is to lie on a plane that has an angle *greater* than $45°$ with the z-axis (such as intersects (3) on a space-like geodesic). On the other hand, from (3),

$$y \leqslant +\sqrt{(R^2 + z^2)}.$$

Hence y too has an upper boundary, therefore $y + z$ has and, from the last (48), $\bar{t}$ also. This was to be proved.

The stopping dead of all peculiar motion including that of light-signals is not a peculiar feature of this Lemaître frame. We shall investigate it later on for flat or spherical universes expanding or contracting according to an arbitrary time law. It has *inter alia* the effect that a particular observer cannot be reached by signals of any kind, from large portions of the world of which he has formed a picture, because they stop dead before they reach him. On the other hand (just as in the static frame), he will be reached by signals that originated outside his world (from events where $y + z \leqslant 0$). Will they embarrass him? Let us speak of light-signals, since for them we can use our previous computations of parallaxes. But first I wish to remove a little paradox. From (48) our present, space coordinates $\bar{x}, \bar{u}, \bar{v}$ are in a very simple relation to the x, u, v, that have been used as space coordinates in the static frame. In the latter the orbits of light-rays were straight, those of particles were not. How comes it that both light-rays and particle orbits are straight in $\bar{x}, \bar{u}, \bar{v}$, but the latter are not straight in the static frame? Or, the other way round, why is the straightness of light-rays conserved in the static frame, while that of particle orbits is not? The first is rather amazing, because the denominator $y + z$ is by no means a constant, and the linear equations between $\bar{x}, \bar{u}, \bar{v}$ are not homogeneous.

What is an orbit? It is the succession of points in contemporary space that carry the same spatial labels as the points passed by the movable object carried. This orbit is straight for a light-signal in $\bar{x}, \bar{u}, \bar{v}$. But time $\bar{t}$ was advancing while the orbit was traversed. The relation between the barred and the unbarred variables depends on time. Why should the orbit be straight also in x, u, v? Well, the (x, u, v)-space is *expanding* uniformly (with respect to the barred frame) at exactly the same rate as that at which the velocity of light in the barred frame decreases. Or better, the $\bar{x}$-frame is contracting with respect to the x-frame. Hence a movable that would proceed with *constant*

91

'$\bar{t}$-velocity' on a straight line in the x-frame would give the correct, motion in the $\bar{x}$-frame and vice versa. Now the t-velocity in the x-frame is not constant and we have not investigated its $\bar{t}$-velocity and we do not, wish to do so, it is irrelevant. For with the x-frame *static* the time rate of advance makes no difference to the orbit *there*. We also understand that the 'preservation of straightness' is confined to light-rays, and does not hold for particle orbits in the static frame.

Now we return to the question how light-signals originating from outside the region covered by this frame, i.e. from $y + z < 0$, will be interpreted in this frame, and whether they will cause embarrassment, in this frame. Now this includes the region of events E for which we have investigated the same question in the static frame; the central observer also may be taken to be the same in both cases. So we establish again two observers there at a small distance, for observing parallaxes. *If we put them*

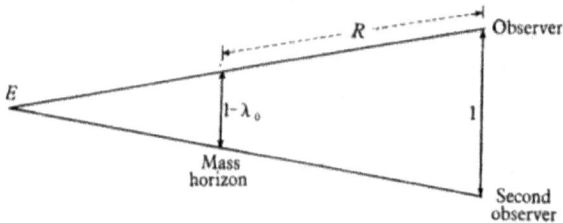

Fig. 5 Showing schematically the parallax that two observers, both at rest near the centre of the static model, find for an extramundane event. The displacement of the observers is taken to be unity, the displacement of their observations, which they project to the mass horizon, is then $1 - \lambda_0$. It is positive but smaller than 1, since the same holds for λ_0. The parallax (for unit displacement of the observers) is λ_0/R. In the figure it is assumed that they locate the event at E, thus at the distance $(1 - \lambda_0) R/\lambda_0$ behind the mass horizon (by straight continuation of their light-rays, which they know to be straight). The objection against this procedure is discussed in the text.

92

in exactly the same conditions as before, the actual setting of their telescopes must be the same, because that cannot depend on the frame which they use for interpreting their observations. We found that they measured a positive parallax (converging rays of vision) such as corresponded, for a displacement 1 of the observers, to a displacement $1 - \lambda_0$ at a coordinate distance R; the displacements referred to (x, u, v)-space, and λ_0 was a number between *zero* and *one*, *zero* for E very far outside their world, *one* for E on their horizon. (Remember the embarrassment, they found even in spatially locating the event; they could not retain beyond the horizon *both* the law of straight propagation, known to them in (x, u, v)-space within, *and* the spherical character of space, known to them from their hemisphere.) Now here, at first sight, there *seems* to be no difficulty. We must take it that they know by experiments (i) that light is propagated along straight lines in $(\bar{x}, \bar{u}, \bar{v})$, (ii) that *this space is* flat and therefore infinite, while *we* know (iii) that

$$\bar{x} = \frac{Rx}{y + z}, \text{etc.}, \tag{57}$$

hence that they must find *ceteris paribus* the parallax calculated before and indicated above. It seems that they would have no difficulty in locating E and, after this, computing a real value of $\bar{t}$ for it. The only thing that may amaze us is that, if this consideration were correct, they would *never* come to locate an event outside the distance R (in $\bar{x}, \bar{u}, \bar{v}$) *except* those events that happen outside their world. They would be wrong. For it is easy to see that they can get light-signals from any distance (in the barred space) provided the signals are launched sufficiently early. So there must be a hitch. What is it?

Well, we must assume that the two observatories are at a constant *invariant* distance from each other, for two reasons: first for their convenience, since their observatories are built in the same country on solid ground, which does not give way; they are certairly *not* built on floating islands; but mainly because, in order to carry over our previous result concerning the parallax, we must put them in exactly the same conditions as the two observers in static space which were of course at fixed places in (x, u, v)-space. From (52), applied to the finite but small interval between them, say $\Delta\bar{x}, \Delta\bar{u}, \Delta\bar{v}$, we see that these displacements must change proportionally to $e^{-\bar{t}}$. Or from (57), which may be written $\bar{x} = xe^{-\bar{t}}$ etc., we have that, for constant, x etc.,

$$dx = 0 = e^{-\bar{t}} + \bar{x}e^{-\bar{t}}d\bar{t} \text{ etc;}$$

hence if the main observer is at rest at $\bar{x} = \bar{u} = \bar{v} = 0,$, the auxiliary observer moves (in the barred space) with velocities

$$\frac{d\bar{x}}{Rd\bar{t}} = -\frac{\Delta\bar{x}}{R}, \text{ etc.,} \tag{58}$$

which means in the direction *towards* the main observer. Knowing that straight propagation holds in the barred space in which he is in motion, he will take into account astronomical aberration, due to his motion. The direction of the motion is such as to feign a positive parallax. If at the moment of observation the displacement, between the observers, orthogonal to the direction of the telescope, is $\Delta\bar{r}$, and $\bar{t}$ is the time, then since from (52) the velocity of light is $e^{-\bar{t}}$ the duly inferred *feigned, parallax* is

$$\frac{\Delta\bar{r}e^{\bar{t}}}{R}.$$

The measured parallax can be gathered from Fig. 5, which refers to (x, u, v)-space; hence the small pieces across (called 1 and $1 - \lambda_0$ there for simplicity) must, be multiplied by Δr, where

$$\Delta r = \Delta\bar{r}e^{\bar{t}}.$$

Thus the *measured parallax* is

$$\frac{\lambda_0 \Delta r}{R} = \frac{\lambda_0 \Delta \bar{r} e^{\bar{t}}}{R}.$$ (59)

Subtracting the feigned parallax, duly suspected by our observer, we obtain

$$-\frac{(1 - \lambda_0)\Delta \bar{r} e^{\bar{t}}}{R} \leqslant 0.$$ (60)

The same result would be obtained by letting the *static* parallax observer more away from the main observer in such a way that his barred coordinates are constant. Then he would find the negative parallax (60), but would calmly correct it to (59) for *his* aberration.

For events very far outside, λ_0 approaches to zero, the negative parallax has its greatest absolute value. It becomes zero for $\lambda_0 \to 1$, that is when E is just on the border of the world. No reasonable interpretation is possible. We must take it that the light-rays actually *converge* in the barred frame. That is puzzling, because they could not really meet again! But the observers can compute from (60) the distance a behind their backs where they meet,

$$\frac{\Delta \bar{r}}{a} = \frac{(1 - \lambda_0)\Delta \bar{r} e^{\bar{t}}}{R}$$

$$a = \frac{R e^{-\bar{t}}}{1 - \lambda_0}.$$

But, would they get there? How far would they get?

$$\int_{\bar{t}}^{\infty} e^{-\bar{t}} R d\bar{t} = R e^{-\bar{t}} \leqslant a,$$

the equality sign holding only in the limit $\lambda_0 \to 0$, i.e. when they come from an event E *very* far outside the frame.

95

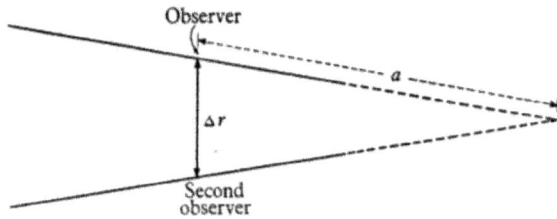

Fig. 6 Illustrating how tho observers, using the Lemaïtre-Robertson frame compute from the negative parallax (converging light-rays) the distance a at which the rays should meet behind their backs. However, they can be satisfied that the rays never reach the point where they would cross (see text).

II. The Theory of Geodesics

7. On null geodesics

A geodesic may be defined as giving a stationary value to the line integral of the interval, for fixed limits,

$$\delta \int ds = 0, \qquad ds^2 = g_{ik}\, dx_i dx_k. \tag{1}$$

It is easy to show that this leads to the Euler equations

$$\frac{d^2 x_k}{ds^2} + \left\{ \begin{matrix} k \\ l\, m \end{matrix} \right\} \frac{dx_l}{ds}\frac{dx_m}{ds} = 0, \tag{2}$$

if after performing the variation with an arbitrary parameter one identifies the parameter with s. (With another parameter (2) is in general modified, see below.) Equations (2) have the first integral

$$g_{ik}\frac{dx_i}{ds}\frac{dx_k}{ds} = \text{constant}. \tag{3}$$

From the second equation (1) the constant is 1. (That the constant from (2) is not determined is connected with the fact that (2) reads alike for any parameter that is a *constant* multiple of s.) It follows that the first member of (3) and therefore the second member of the second equation (1) can vanish nowhere on our curve.

From general theory the solutions of (2) exist for any initial values of $x_k, dx_k/ds$, and (3) is always a first integral.

This holds irrespective of the meaning of the parameter s. Thus if the initial values are chosen so as to make the first member of (3) vanish, it would vanish all along the curve, *for any admissible choice of the parameter,* say τ. Then from (1)

$$\left(\frac{ds}{d\tau}\right)^2 = 0$$

all along the curve, showing that s is in this case not an admissible parameter. But worse than that, the variational principle (1) is in this case meaningless with *any* parameter. It reads more explicitly

$$\delta \int \sqrt{\left(g_{ik}\frac{dx_i}{d\tau}\frac{dx_k}{d\tau}\right)}\, d\tau = 0. \tag{4}$$

Now $\sqrt{x}$ has all its derivatives infinite at, $x = 0$ and cannot be developed at that point with respect to a small increment of the radicand.[1] Hence wherever the square root, vanishes the variation (4) cannot, be performed. Moreover, for some small increments the integrand would be real, for others imaginary, which is inadmissible in the calculus of variations.

We therefore make quite a general change in the variational principle, to define geodesics. We adopt

$$\delta \int g_{ik}\frac{dx_i}{d\tau}\frac{dx_k}{d\tau}\, d\tau = 0. \tag{5}$$

At first sight one may think that its extremals in the non-singular case are different from those of (1). But this is not so. Without computing them, we conclude from the complete analogy with analytical mechanics that they admit

---

[1]Of course this does not, mean that it changes by an infinite amount, only by an amount not linear in the increment, but proportional to its square root.

the first integral

$$g_{ik}\frac{dx_i}{d\tau}\frac{dx_k}{d\tau} = \text{constant} = E \quad \text{(say)}. \qquad (6)$$

This constant may or may not be zero. If it is not zero, you see that $d\tau$ is, apart from a multiplying constant, the same as ds. It is a, very remarkable fact that our variational principle (5), posited first in an arbitrary parameter, yet determines that parameter up to a linear transformation with constant coefficients. Proved, for the time being, for a non-vanishing constant in (6). Now it is small wonder that for $E \neq 0$ (5) has the same extremals as (1). Indeed, you may in this case write for (5):

$$\delta \int \left[\sqrt{\left(g_{ik}\frac{dx_i}{d\tau}\frac{dx_k}{d\tau} \right)} \right]^2 d\tau$$

$$= 2 \int \sqrt{E}\, \delta \sqrt{\left(g_{ik}\frac{dx_i}{d\tau}\frac{dx_k}{d\tau} \right)}\, d\tau$$

$$= 2\sqrt{E}\, \delta \int \sqrt{\left(g_{ik}\frac{dx_i}{d\tau}\frac{dx_k}{d\tau} \right)}\, d\tau.$$

This does *not* show the equivalence for $E = 0$. But in this case *there is no equivalence;* for (5) remains meaningful, while (1) does not. Following Levi-Civita,[2] we *extend*, the definition of geodesics to null geodesics by (5). It is very easy to show that the differential equations of the extremals of (5) read in every case

$$\frac{d^2 x_k}{d\tau^2} + \left\{ \begin{matrix} k \\ l\,m \end{matrix} \right\} \frac{dx_l}{d\tau}\frac{dx_m}{d\tau} = 0. \qquad (7)$$

This, together with (6) for $E = 0$, defines the null geodesics. Of course in this case $d\tau$ is now *not ds* (which is zero). At

---

[2] *The Absolute Differential Calculus* (Blackie and Son, 1929), p. 332.

first sight one would think that τ now remains undetermined.[3] But this is not so. For if you introduce in (7) $\lambda = \lambda(\tau)$, you find

$$\frac{d^2 x_k}{d\lambda^2} + \left\{ \begin{matrix} k \\ l\, m \end{matrix} \right\} \frac{dx_l}{d\lambda} \frac{dx_m}{d\lambda} = -\frac{\lambda''}{\lambda'^2} \frac{dx_k}{d\lambda}. \qquad (7\text{a})$$

Hence the simpler form (7) can hold only for a particular parameter τ distinguished up to a linear transformation with constant coefficients. For non-null geodesics it is essentially s, the length of the arc. This stipulation cannot, be extended to null geodesics. But another one can, viz. that $dx_k/d\tau$ is from (7) *the tangent vector parallel-transferred, along the curve.* In the non-singular case you need not actually transfer it; it is easily recognized because it is at any point the one that has the same invariant, length as anywhere else. In the singular case this test fails, because *all* tangent vectors have the length zero. This is inconvenient. Still you may freely use (7) – and need not resort to (7a) – either for determining a null geodesic from appropriately chosen initial conditions, or for determining the geodesic between two given points, *at the risk* that it turns out, a null geodesic. For you know that the parameter τ in which this simple form of the equations obtains exists in any case.

(a) Determination of the parameter for null lines in special cases

In some cases the appropriate differential $d\tau$ can be made out easily, viz. when at least one of the four coordinates is 'cyclic,' by which we mean that the g_{ik} do not depend on it.[4] The Euler equations of a cyclic variable can be immediately integrated and define $d\tau$. The most important case

---

[3]Even good recent text-books succumb to this error.

[4]The name 'cyclic' is given in analytical mechanics to a variable, when the energy depends only on the time derivative of that variable,

is that of a *static* gravitational field: the g_{ik} independent of x_4 and $g_{ik} = 0(k = 1, 2, 3)$. (Even without the second assumption x_4 is cyclic. But the case of a *stationary* field is less simple and not so important in application.)

Let us write Q for the integrand of (5), so that (5) becomes

$$\delta \int Q d\tau = 0. \tag{8}$$

In our present case

$$Q = \sum_1^3 g_{ik} \frac{dx_i}{d\tau} \frac{dx_k}{d\tau} + g_{44} \frac{dx_4}{d\tau} \frac{dx_4}{d\tau}.$$

The Euler equation from varying x_4 reads

$$\frac{d}{d\tau}\left(g_{44}\frac{dx_4}{d\tau}\right) = 0, \qquad g_{44}\frac{dx_4}{d\tau} = 1; \tag{9}$$

really only: $=$constant. But the constant cannot vanish, since neither factor on the left can, and so by an irrelevant change in $d\tau$ we can have it 1. Hence

$$d\tau = g_{44} dx_4 \tag{10}$$

is an appropriate choice of the parameter. In other words the tangential vector

$$\frac{1}{g_{44}}\frac{dx_1}{dx_4}, \quad \frac{1}{g_{44}}\frac{dx_2}{dx_4}, \quad \frac{1}{g_{44}}\frac{dx_3}{dx_4}, \quad \frac{1}{g_{44}} \tag{11}$$

is parallel-transferred along a null geodesic in a static gravitational field. (It is a vector for all significant transformations that leave the field static, viz. transformations of the first three coordinates.)

Allow me a short digression (on the principle of Fermat and its analogue in mechanics). If we compute the variation (5) *for a null geodesic* in a static field, but allowing

---

not on its undifferentiated value.

101

x_4 to vary at the end-points, it is not generally zero, but we get a contribution from the limits (which we indicate by '1' and '2'):

$$\delta \int Q d\tau = g_{44} \frac{dx_4}{d\tau} \delta x_4 \Big|_1^2 .$$

If, however, we apply this only to such 'varied paths' on which Q is also zero, i.e. which consist of null elements without being null *geodesics*, then the variation is zero, and we obtain, using (9),

$$0 = g_{44} \frac{dx_4}{d\tau} \delta x_4 \Big|_1^2 = \delta x_4 \Big|_1^2 = \delta \int_1^2 dx_4 .$$

This is Fermat's principle of least (or stationary) time of transition. It holds precisely under the condition of variation which we have adopted, viz. that the comparison curves join the same two points in *space* and be followed throughout 'with the local velocity of light', i.e. $Q = 0$. It can obviously be put in the three-dimensional form

$$\delta \int_1^2 g_{44}^{-1/2} \sqrt{\left(-\sum_1^3 g_{ik} dx_i dx_k \right)} = 0.$$

The spatial end-points '1' and '2' are fixed. The parameter to be used is unprejudiced. The relation between the four-dimensional geodesic principle and Fermat's principle is analogous to that of Lagrange and that of Maupertuis in general mechanics. According to the former

$$\delta I = \delta \int_{t_1}^{t_2} (T - V) dt = 0$$

for fixed limits, i.e. this gives the equations of motion. If the system is conservative

$$T + V = E \quad \text{is a first integral.}$$

If one varies also t, the time of transition, then

$$\delta I = -E\delta t \Big|_{t_1}^{t_2} \left(not \text{ perhaps } (T-V)\,\delta t \Big|_{t_1}^{t_2} \right).$$

(We defer the proof for a moment.) Now let

$$M = \int_{t_2}^{t_2} 2T\,dt = \int_{t_1}^{t_2} (T-V)\,dt + \int_{t_1}^{t_2} (T+V)\,dt.$$

If we now perform a variation in which we keep E constant and allow the limits to vary, then

$$\delta M = \delta \int_{t_1}^{t_2} 2T\,dt = -E\delta t \Big|_{t_1}^{t_2} + E\delta t \Big|_{t_1}^{t_2} = 0.$$

This is Maupertuis's principle. The integral can be put in a form where it no longer contains $t : 2T = 2(E-V)$, and for T we write

$$\frac{1}{2}\left(\frac{ds}{dt}\right)^2, \quad \text{so} \quad dt = \frac{ds}{\sqrt{(2T)}},$$

thus

$$\delta \int_{B}^{A} ds\sqrt{\{2(E-V)\}} = 0.$$

Notice that here it is *not* the time of transition that is a minimum!

To prove the expression used for δI directly, introduce a parameter λ, which does *not* vary at the limits and take t to be a function of λ; the derivatives with respect to λ. are indicated by a dot:

$$dt = \dot{t}\,d\lambda, \quad T = \frac{1}{2}\left(\frac{ds}{dt}\right)^2 = \tfrac{1}{2}\dot{s}^2\dot{t}^{-2},$$

103

$$\delta I = \delta \int_1^2 \left(\tfrac{1}{2} \dot{s}^2 \dot{t}^{-2} - V \right) \dot{t} d\lambda$$

$$= \int_1^2 \left(-\tfrac{1}{2} \dot{t}^{-2} \dot{s}^2 \dot{t}^{-2} \delta \dot{t} - V \delta \dot{t} \right) d\lambda,$$

since only the contribution from varying $t(\lambda)$ is relevant, the rest vanishing from Lagrange's principle. Hence

$$\delta I = \int_1^2 \frac{d}{d\lambda} \left(\tfrac{1}{2} \dot{s}^2 \dot{t}^{-2} + V \right) \delta t d\lambda - \left| \left(\tfrac{1}{2} \dot{s}^2 \dot{t}^{-2} + V \right) \delta t \right|_1^2$$

$$= -E \delta t \Big|_{t_1}^{t_2}$$

because the integrand vanishes. (This result is not confined to conservative systems, but in the general case it is not so simple to prove.) After this digression we return to our main topic. A non-static case of interest is the de Sitter line element in the Lemaïtre frame:

$$Q = -e^{2\bar{t}} \left[\left(\frac{d\bar{x}}{d\tau} \right)^2 + \left(\frac{d\bar{u}}{d\tau} \right)^2 + \left(\frac{d\bar{v}}{d\tau} \right)^2 \right] + R^2 \left(\frac{d\bar{t}}{d\tau} \right)^2.$$

$$\tag{12}$$

Here all three space coordinates are cyclic. At least one of them must vary along the ray. Its Euler equation can be integrated, and gives

$$e^{2\bar{t}} \frac{d\bar{x}}{d\tau} = 1, \quad d\tau = e^{2\bar{t}} d\bar{x}.$$

The parallel-transferred tangential vector is

$$e^{-2\bar{t}}, \quad e^{-2\bar{t}} \frac{d\bar{u}}{d\bar{x}}, \quad e^{-2\bar{v}} d\bar{x}, \quad e^{-2\bar{t}} \frac{R d\bar{t}}{d\bar{x}}.$$

But since this is not a vector for orthogonal transformations of the three space coordinates (the only significant ones that leave (12) unchanged), it is better to take

$$d\tau = e^{2\bar{t}} d\bar{\tau} (d\bar{\sigma}^2) \quad (d\bar{\sigma}^2 = d\bar{x}^2 + d\bar{u}^2 + d\bar{v}^2)$$

104

and get the parallel-transferred tangential vector

$$e^{-2\bar{t}}\frac{d\bar{x}}{d\bar{\sigma}}, \quad e^{-2\bar{t}}\frac{d\bar{u}}{d\bar{\sigma}}, \quad e^{-2\bar{t}}\frac{d\bar{v}}{d\bar{\sigma}}, \quad e^{-\bar{t}}$$

(because $R d\bar{t}/d\bar{\sigma} = e^{\bar{t}}$ on the null line). The appropriate parameter is *not* the invariant spatial *length* along the ray, that is, $e^{\bar{t}}d\bar{\sigma}$, and the same as $R d\bar{t}$. So one may also take $\tau = e^{\bar{t}}$.

(b) Frequency shift

The knowledge of the appropriate parameter on null lines, when it, can be procured, is helpful in determining the frequency shift of light emitted along one time-like worldline and observed on another one. There would be in general no point in comparing the two frequencies measured in world time, because they depend on the frame. Furthermore, in the relevant cases, we can only observe one of them, the source being inaccessible (on the sun or on a star or nebula). What we actually compare is the frequency of the light coming from an external source, measured in our eigen-time, and the frequency, measured on the same scale, of the light of the same light-producer (e.g. a sodium atom) in our laboratory. The assumption is that the latter frequency is an invariant, thus the same as at the distant, source, measured in the eigen-time of the source. Therefore, to predict the observed shift from theory, we must determine the ratio of two intervals of eigen-time, δs_1 at the source between the emission of two successive pulses of light (wave-crests, for instance), and δs_2 at the receiver between the arrival of these pulses. The ratio

$$\frac{\delta s_2}{\delta s_1} \tag{13}$$

determines the shift: the *period* we observe in the light coming from the distant source, divided by the period an observer *at* the source would observe, that is, by assump-

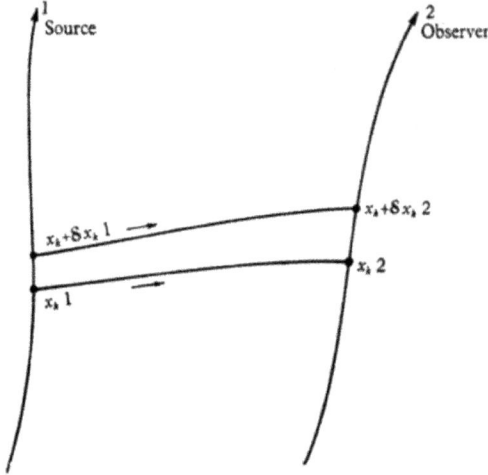

Fig. 7 Illustrating the derivation of the general formula (16) for the frequency shift from the fact that the Lagrangian integral is zero for every light-ray and must therefore, for two subsequent light pulses traveling from tho source to the observer, be the same even though both limits of the integral change according to the motion of the source and that of the observer, which may both be quite arbitrary.

tion, the same as we measure for the same light-producer in the laboratory. Thus the above ratio, when it is greater than 1, means a red shift, when smaller a violet shift. It is called a Doppler effect and/or a gravitational effect. But while the total shift itself is invariant, the two kinds of shift cannot, be distinguished invariantly.

To determine the ratio (13) we compute the variation of the integral (5), the unvaried path of integration being the null line (light-ray) from some point on the world line of the source, $x_k|_1$, to the point $x_k|_2$, on the world line of the observer, while the varied path shall be the null line from $(x_k+\delta x_k)|_1$ to $(x_k+\delta x_k)|_2$. We must pay attention to the variations at the limits. The total variation vanishes, of course, because the integral vanishes along both paths. So we have

$$0 = \delta \int_1^2 g_{ik} \frac{dx_i}{d\tau} \frac{dx_k}{d\tau} d\tau$$

$$= \int_1^2 \left(2g_{ik} \frac{dx_i}{d\tau} \frac{d\delta x_k}{d\tau} + g_{ik,l} \frac{dx_i}{d\tau} \frac{dx_k}{d\tau} \delta x_1 \right) d\tau$$

$$= 2g_{ik} \frac{dx_i}{d\tau} \delta x_k \Big|_1^2 -$$

$$- \int_1^2 \left[2\frac{d}{d\tau} \left(g_{ik} \frac{dx_i}{d\tau} \right) - g_{im,k} \frac{dx_i}{d\tau} \frac{dx_m}{d\tau} \right] \delta x_k d\tau. \qquad (14)$$

The integral vanishes, by (7), *for the appropriate parameter τ*. Hence for *it*

$$g_{ik} \frac{dx_i}{d\tau} \delta x_k \Big|_1^2 = 0 \qquad (15)$$

Therefore

$$\frac{\delta s_2}{\delta s_1} = \frac{\left[g_{ik} \frac{dx_i}{d\tau} \frac{\delta x_k}{\delta s} \right]_1}{\left[g_{ik} \frac{dx_i}{d\tau} \frac{\delta x_k}{\delta s} \right]_2} \qquad (16)$$

The bracketed expression is the scalar product of the tangent vectors of the world line of motion (δ) and of the light-ray (d), at the source (1) and the receiver (2), respectively. These products cannot vanish, because a time-like vector cannot be orthogonal to a non-vanishing null vector.[5] It is

---

[5] It will suffice to verify this after transforming g_{ik} to the metrical tensor of special relativity (-1, -1, -1, 1). Let p_k be a null vector, thus $p_k p^k = 0$. Let u_k be orthogonal to p_k thus $p_k u^k = 0$. On account of these two relations the norm of u_k is, for *any* λ the same as that of $u_k + \lambda p_k$. The norm of the latter reads *in extenso*

$$-(u_1 + \lambda p_1)^2 - (u_2 + \lambda p_2)^2 - (u_3 + \lambda p_3)^2 + (u_4 + \lambda p_4)^2.$$

By taking tho special value $\lambda = -u_4/p_4$ it is seen that this norm cannot be positive and vanishes only when u_k is a multiple of p_k. In other words the vectors orthogonal to a null vector, except for its own multiples, are all space-like.

clear that the field g_{ik} must, be known at both places and so must the four directions, all this explicitly, if the question about the frequency shift is to have a definite answer. The only thing that is missing for evaluating the second member of (16) is the ratio of $d\tau$ at (1) and $d\tau$ at (2), in other words the parameter τ for which the integrand in (14) vanishes. From this vanishing we know that $\frac{dx_i}{d\tau}\big|_2$ is $\frac{dx_i}{d\tau}\big|_1$ parallel-transferred along the ray. This must be found out in every special case unless it is of the simple kind, mentioned before, where it is known in general.

An alternative to transferring the light vector (the d-quotients) is to transfer from (1) to (2) the tangent vector of the time-like world line (the δ-quotients). For the numerator remains unchanged when all its three factors are transferred; when transferred the first two become identical with the first two of the denominator and we may say that the frequency shift is determined as the quotient of the cosines of the two angles which the light-ray subtends with the direction vector of the source (transferred to the receiver), and with the direction vector of the receiver, respectively.

Let us test (16) in the simplest case of a static field with both the source and the receiving observer *at rest*. This leaves at both limits only $\delta x_4/\delta s \neq 0$ and equal to $g_{44}^{-1/2}$. Both in the numerator and in the denominator only the term $i = k = 4$ survives. The appropriate $d\tau$ by (10), is $g_{44}dx_4$. This gives $dx_4/d\tau = 1/g_{44}$. So we obtain

$$\frac{\delta s_2}{\delta s_1} = \frac{\left[g_{44}\frac{1}{g_{44}}g_{44}^{-\frac{1}{2}} \right]_1}{\left[g_{44}\frac{1}{g_{44}}g_{44}^{-\frac{1}{2}} \right]_2} = \frac{\sqrt{g_{44}}\,|_2}{\sqrt{g_{44}}\,|_1}$$

This agrees with the well-known 'gravitational red shift'. It can also be obtained by the simple consideration that the observed *wave-lengths* must be in direct proportion to the *light-velocities* (in world time!), i.e. to the $\sqrt{g_{44}}$ since all emitted pulses reach the receiver and take the same world time to reach him; a fact which one must be careful not, to interpret as meaning that the *frequency* is unchanged: for the frequency in world time has no observational meaning.

Let us now look at the Lemaître-Robertson frame

$$ds^2 = -e^{2\bar{t}}(d\bar{x}^2 + d\bar{u}^2 + d\bar{v}^2) + R^2 d\bar{t}^2$$

for which we also write

$$= -e^{-2\bar{t}}d\bar{\sigma}^2 + R^2 d\bar{t}^2.$$

We wish to determine the frequency shift when both source and observer are *at rest in the barred, coordinates* $\bar{x}$, etc. The observer may be placed at the spatial origin. We found the appropriate $d\tau$ to be $e^{2\bar{t}}d\bar{\sigma}$. What do we get from (16)? At both ends, only

$$\frac{\delta x_4}{\delta s} = \frac{\delta \bar{t}}{\delta s} = \frac{1}{R}$$

differs from zero. So we need for the light-ray only

$$\frac{dx_4}{d\tau} = \frac{d\bar{t}}{e^{2\bar{t}}d\bar{\sigma}} = e^{-2\bar{t}}.e^{\bar{t}}R^{-1} = e^{-\bar{t}}\frac{1}{R} \qquad \text{(since } ds = 0\text{).}$$

Hence from (16)

$$\frac{\delta s_2}{\delta s_1} = \frac{R^2 e^{-\bar{t}_1} R^{-2}}{R^2 e^{-\bar{t}_2} R^{-2}} = e^{\bar{t}_2 - \bar{t}_1} > 1,$$

meaning a red shift. (For remember, this is the ratio of the periods or wave-lengths.) Let $\bar{r} = \sqrt{(\bar{x}^2 + \bar{u}^2 + \bar{v}^2)}$ be the 'barred' distance which, by assumption, does *not* change. Then from the line element (for $d\bar{t} = 0$ the *invariant* distance, *which does change*, is obviously $\bar{r}e^{\bar{t}}$. So

$$\frac{\lambda_2}{\lambda_1} = \frac{\text{invariant distance at reception (2)}}{\text{invariant distance at emission (1)}} = \frac{r_2}{r_1} \qquad \text{(say).}$$

At first sight one is mildly astonished. From the naïve point of view the observer is at rest and the source moves outward at the rate $e^{\bar{t}_1}$ at the moment of emission; should not this alone determine the frequency shift? This naive idea is not, far out, but in this form only approximate. Write

$$\frac{\lambda_2}{\lambda_1} = \frac{\lambda_1 + \Delta\lambda}{\lambda_1} = 1 + \frac{\Delta\lambda}{\lambda_1} = \frac{r_2}{r_1}$$

hence

$$\frac{\Delta\lambda}{\lambda_1} = \frac{r_2 - r_1}{r_1}.$$

Now $r_2, -r_1$ is the distance covered by the source between $\bar{t}_1$ an $\bar{t}_2$ and r_1 the distance covered by the light in the same time, and thus, if you disregard the acceleration of the source, this fraction is to the naive view very nearly the ratio of their velocities (which is the familiar approximation to the value of $\delta\lambda/\lambda$).

But, our result purports to be exact. So let us check it in a more direct way. Let us envisage again the two subsequent light-pulses traveling from the world line of the source (1) to that of the observer (2). Now let us measure the distance $\bar{r}$ from A_1 to A_2 along the straight light ray. From the line element, and the null property

$$d\bar{r} = R\,e^{-\bar{t}}d\bar{t},$$
$$\bar{r} = R\left(e^{-\bar{t}_1} - e^{-\bar{t}_2}\right).$$

By assumption it does not change and must be the same as from B_1 to B_2. Hence

$$-e^{-\bar{t}_1}\delta\bar{t}_1 + e^{-\bar{t}_2}\delta\bar{t}_2 = 0,$$
$$\frac{\delta\bar{t}_2}{\delta\bar{t}_1} = e^{\bar{t}_1-\bar{t}_2} = \frac{\delta s_2}{\delta s_1},$$

because both are at rest in the barred frame. This is the

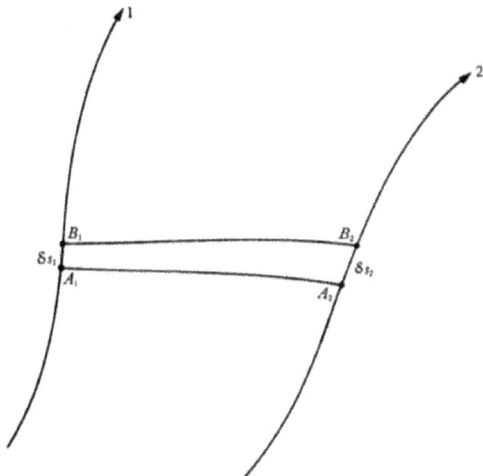

Fig. 8 Illustrating the way in which the ratio ds_2/ds_1 (and thereby the red shift) in tho Lemaître-Robertson frame, with sources and observer both at rest in this frame, is computed directly from the fact that the coordinate distance between them does not change with time.

same result as before. But we may now put it into a different form, using the above result for $\bar{r}$; from it

$$\frac{\bar{r}e^{\bar{t}_1}}{R} = 1 - e^{\bar{t}_1 - \bar{t}_2} = 1 - \frac{\delta s_1}{\delta s_2} = 1 - \frac{\nu_2}{\nu_1}$$

or in an obvious notation

$$-\frac{\Delta\nu}{\nu_1} = \frac{\bar{r}e^{\bar{t}_1}}{R}.$$

This conforms *exactly* to the naïve view, since the second member is just the ratio of what, on this view is the velocity of the source on emission and 'the' velocity of light. (But this would not hold for other laws of expansion.)

111

8. Free particles and light rays in general, expanding spaces, flat or hyperspherical

(a) Flat spaces

We generalize the Lemaître line element (52) (Chapter 1) as follows:

$$ds^2 = -e^{2g(\bar{t})}(d\bar{x}^2 + d\bar{u}^2 + d\bar{v}^2) + R^2 d\bar{t}^2, \qquad (17)$$

where $g(\bar{t})$ is an arbitrary function of $\bar{t}$, while R is some constant length, which we retain for the sake of comparison with §6, though it has now no obvious meaning. Of course (17) is no longer a solution of the field equations in an empty world but needs a (spatially uniform) density, and, maybe, pressure to support it. We do not propose to investigate them here; the general problem is expounded in great detail in R. C. Tolman's book.[6] We take (17) as given and wish to study its geodesics, more particularly the time-like and null ones. We use the variational principle (5) of §7, which is applicable to all cases,

$$\delta \int \left[-e^{2g(\bar{t})} \left\{ \left(\frac{d\bar{x}}{d\tau}\right)^2 + \left(\frac{d\bar{u}}{d\tau}\right)^2 + \left(\frac{d\bar{v}}{d\tau}\right)^2 \right\} \right.$$
$$\left. + R^2 \left(\frac{d\bar{t}}{d\tau}\right)^2 \right] d\tau = 0. \qquad (18)$$

We know the first integral (6), §7,

$$-e^{2g(\bar{t})} \left\{ \left(\frac{d\bar{x}}{d\tau}\right)^2 + \left(\frac{d\bar{u}}{d\tau}\right)^2 + \left(\frac{d\bar{v}}{d\tau}\right)^2 \right\} + R^2 \left(\frac{d\bar{t}}{d\tau}\right)^2 = E,$$
$$\qquad (19)$$

a constant which is zero for null geodesics, positive for time-like. A little consideration shows that, on varying

---

[6] *Relativity, Thermodynamics and Cosmology*, pp. 361 ff.

the spatial parameter $\bar{x}, \bar{u}, \bar{v}$ we get three very simple Euler equations, e.g

$$\frac{d}{d\tau}\left(e^{2g(\bar{t})}\frac{d\bar{x}}{d\tau}\right) = 0, \tag{20}$$

formally the same as for the three-dimensional geodesic problem of static flat *space*, with which our set, (20) coincides if the parameter

$$d\tau' = e^{-2g(\bar{t})}d\tau \tag{21}$$

is used. We conclude

(i) that *all orbits are straight lines in the barred, space coordinates*, but, may degenerate into one *point* (because neither $d\tau$ nor $d\tau'$ is the spatial line element);

(ii) that we have also the first integral of the three-dimensional problem (easily confirmed from the three equations like (20))

$$e^{4g(\bar{t})}\left\{\left(\frac{d\bar{x}}{d\tau}\right)^2 + \left(\frac{d\bar{v}}{d\tau}\right)^2 + \left(\frac{d\bar{v}}{d\tau}\right)^2\right\} = f^2, \tag{22}$$

a non-negative constant (for $f = 0$ we get the point mentioned under (i)). By carrying (22) into (19) we find the meaning of the parameter τ

$$d\tau = \frac{R d\bar{t}}{\sqrt{E + f^2\, e^{-2g(\bar{t})}}}. \tag{23}$$

Hence from (22), the 'barred' velocity

$$\overline{V} = \sqrt{\left(\frac{d\bar{x}}{d\bar{t}}\right)^2 + \left(\frac{d\bar{u}}{d\bar{t}}\right)^2 + \left(\frac{d\bar{v}}{d\bar{t}}\right)^2} = \frac{R f\, e^{-2g(\bar{t})}}{\sqrt{E + f^2\, e^{-2g(\bar{t})}}}. \tag{24}$$

We notice that, just as in the Lemaître-de Sitter case, all particles, including light-signals ($E = 0$), stop dead asymptotically if expansion continues indefinitely, i.e. for

$g(\bar{t}) \to \infty$. This holds in the barred frame. It means that, they eventually just share the common expansion. $\bar{V}$ is the velocity of what the astronomer might call the *peculiar* motion, when the expansive motion is discounted. But *besides*, $\bar{V}$ is expressed in a way that is not very suggestive, because a certain time rate of change of $\bar{x}$, say, means at different epochs a different time rate of change of what we are inclined to call the 'true x coordinate whose change is $e^{g(\bar{t})}$ times greater. In redefining the *peculiar* velocity accordingly we shall also take away the factor R in the second member of (24) from (17) this means that for our present purpose it is better to regard $R\bar{t}$ as the time in lieu of $\bar{t}$. We then get

$$V = R^{-1}\bar{V}e^{g(\bar{t})} = \frac{fe^{-g(\bar{t})}}{\sqrt{E + f^2\,e^{-2g(\bar{t})}}}. \qquad (25)$$

We notice that for $E \neq 0$ the redefined peculiar velocity is still running down, though less rapidly. For light, however, it becomes 1 ($f = 0$ must be excluded for $E = 0$, because this would make our parameter τ meaningless, from (23)).What is the meaning of E when it does not vanish? Compute

$$\sqrt{(1 - V^2)} = \frac{\sqrt{E}}{\sqrt{E + f^2e^{-2g(\bar{t})}}}, \qquad (26)$$

thus

$$\frac{V\sqrt{E}}{\sqrt{1 - V^2}} = fe^{-g(\bar{t})}. \qquad (27)$$

Now from (19) the value of E is only fixed up to a constant multiplier, since we know this to be the case with τ. But we prefer to adopt an invariant, which will make $d\tau$ an invariant differential. Equation (27) suggests taking for E the square of the rest mass; then (27) is the momentum of *the peculiar motion*. The square root in the denominator

114

of (26) is then the energy of the peculiar motion and the parameter τ is, from (23), essentially the time integral of the reciprocal of this energy. We note that the momentum (27) changes inversely as the expansion factor or as the 'linear dimensions' of the model.

We may reasonably doubt whether it has any good physical meaning to speak of momentum and energy of the *peculiar* motion and disregard the common expansive motion. But it is certainly the simplest way of expressing the actual state of affairs. I dare say that flat models other than the Lemaître one have little physical significance. Things become very much simpler for spherical space.

(b) Spherical spaces

We contemplate the line element

$$ds^2 = -e^{2g(t)} R^2 \left[d\chi^2 + \sin^2 \chi (d\theta d\phi^2) \right] + R^2 dt^2, \quad (28)$$

which is a generalization of that very simple form of representing de Sitter space-time that was given by (8) for the reduced model, but not further investigated. Again (28) needs in general a matter tensor to support it, but we are not concerned with it here. As in (18) we use the 'quadratic' variational principle, which for brevity we write

$$\delta \int \left(\frac{ds}{d\tau} \right)^2 d\tau = 0, \quad (29)$$

or, a little more elaborately,

$$\delta \int \left[-e^{2g(t)} R^2 \left(\frac{d\omega}{d\tau} \right)^2 + R^2 \left(\frac{dt}{d\tau} \right)^2 \right] d\tau = 0. \quad (30)$$

The ω is not a variable, but its derivative with respect to t is an 'angular velocity'. As before we have the first integral

$$\left(\frac{ds}{d\tau} \right)^2 = E. \quad (31)$$

115

The variations with respect to χ, θ, ϕ would here be a little more involved, but need not be carried out, because we see that, using again a parameter τ' as in (21), we are faced with the minimum problem of the three-dimensional static hyper-sphere. So we conclude

(i) that all orbits are great circles on the unit hyper-sphere described by χ, θ, ϕ (which play here exactly the role of the 'barred' space in the flat problem), or possibly a point on it, viz. when χ, θ, ϕ, arc constants; these angles are often called co-moving coordinates;

(ii) that we have another first integral, corresponding to (22)

$$e^{4g(t)} \left(\frac{d\omega}{d\tau} \right)^2 = f^2. \tag{32}$$

From the last two equations

$$- e^{-2g(t)} R^2 f^2 + R^2 \left(\frac{dt}{d\tau} \right)^2 = E, \tag{33}$$

and therefore

$$d\tau = \frac{R\,dt}{\sqrt{E + R^2 f^2 e^{-2g(t)}}}. \tag{34}$$

Using this in (32) we get the *velocity*

$$V = Re^{g(t)} \frac{d\omega}{dt} = \frac{f R^2 e^{-g(t)}}{\sqrt{E + f^2 R^2 e^{-2g(t)}}}. \tag{35}$$

I have omitted the attribute 'peculiar'. It would not be precisely wrong, unless it suggested that there was any other velocity, because there is not. True, two bodies that both have $V = 0$ increase their distance on account of the expansion; there is a mutual red-shift etc.; but this 'velocity' pertains neither to the one nor to the other, and not, half and half to each, because on this view a third and fourth body (with $V = 0$) would cause great, embarrassment.

From (35) we see that a material body ($E > 0$) stops dead asymptotically on the great, circle on which, as we have seen, it moves. But for a light-signal ($E = 0$),$V = R$ (a constant!) (For the moment it might have been seemlier to omit in the last, term of the line element (28) the factor R^2 that we kept for reference to our previous work; this would have meant replacing our Rt by t and producing $V = 1$ for tight.)

In spite of the constancy of V for light, it may happen that a light-signal emitted from a body at rest ($V = 0$) towards another body at, rest does not approach the latter and possibly never reaches it. For if ω is their angular distance, the length of the arc of a great circle between them, viz.

$$Re^{g(t)}\omega, \tag{36}$$

may increase more rapidly than the signal advances along it. Indeed, it may be that

$$Re^{g(t)}g'(t)\omega dt > Rdt.$$

For instance, in the de Sitter case (the line element (8)), where $\cosh t$ replaces our exponential, this is so and remains so after

$$\omega \sinh t > 1.$$

This example shows that a body at rest can be *swept* out of the ken of another body at rest by the sheer expansion. We return to (35) and form, similarly to (27),

$$\frac{\frac{V}{R}\sqrt{E}}{\sqrt{1 - V^2/R^2}} = fRe^{-g(t)}. \tag{37}$$

If we take E to be the square of the rest mass,[7] as we may, this is the momentum. It decreases in inverse pro-

---

[7]Perhaps we ought to say 'Put E equal to the square of the rest

portion to the radius of space. In an expanding space *all momenta decrease in this fashion* for bodies acted on by no other forces than gravitation (which originates, of course, from the matter required to support our line element (28)). This simple law has an even simpler interpretation in wave mechanics: all wave-lengths, being inversely proportional to the momenta, simply expand with space. This can be confirmed by a direct wave-mechanical treatment, which also reveals that even a spatially limited wave phenomenon (e.g. a 'wave parcel') 'shares the expansion of space'; but the explanation of the precise meaning of this expression must be deferred, because it is not quite as simple as it looks. The amplitudes decrease in such a way that the total energy of a (nearly) monochromatic wave parcel decreases proportionally to its frequency. (This satisfies the quantum physicist.) All this holds also for light, which in our present treatment would have to be tackled by a not very satisfactory transition to the limit $E \to 0$, $V \to R$. I ought to add that the wave theorems alluded to are not *quite* exact, but, are very close approximations as long as the expansion is not *extremely* rapid, that is to say as long as $R^{-1}(dR/dt)$ is small compared to the frequencies of the waves in question.

For the energy of our body wo obtain from (35) and (37)

$$
\frac{\sqrt{E}}{\sqrt{1 - V^2/R^2}} = e^{-g(t)}\sqrt{Ee^{2g(t)} + f^2 R^2}
$$

$$
= \sqrt{E + f^2 R^2 e^{-2g(t)}}. \tag{38}
$$

a well-known connection, since (37) is the momentum and E the square of the rest energy.

---

mass times R^2,' since the velocity of light is R (a constant!). The reader will forgive this irrelevant inconsistency.

The energy decreases, of course, and its rate of decrease is governed by another interesting relation. For easier survey I shall first give the non-relativistic approximation. According to the kinetic formula for the pressure of an ideal gas ($\frac{1}{3}m_0\overline{v^2}n$ in familiar notation) a single particle in the total volume Ω of space contributes to the pressure

$$[p] = \frac{m_0 v^2}{3\Omega}. \tag{39}$$

We multiply this by the increase $d\Omega$ of Ω during dt

$$[p]\,d\Omega = \frac{m_0 v^2 d\Omega}{3\Omega} = m_0 v^2 g'(t)dt, \tag{40}$$

since Ω increases as the third power of the radius. The particle's kinetic energy we compute from its momentum (37),

$$\frac{1}{2}m_0 v^2 = \frac{f^2 R^2}{2m_0}e^{-2g(t)}. \tag{41}$$

It decreases during dt by

$$-d(\tfrac{1}{2}m_0 v^2) = \frac{f^2 R^2}{m_0}e^{-2g(t)}g'(t)dt = m_0 v^2 g'(t)dt, \tag{42}$$

the same as (40). *The energy loss is such as would, be spent on pressure work against a receding piston, though there is no piston.* To turn (39) into the exact relativistic formula we have to replace mv^2 by the product of momentum and velocity. From (37) and (35) this gives

$$[p] = \frac{1}{3\Omega}\frac{(V/R)^2\sqrt{E}}{\sqrt{1-(V/R)^2}} = \frac{f^2 R^2 e^{-2g(t)}}{3\Omega\sqrt{E + f^2 R^2 e^{-2g(t)}}}, \tag{39'}$$

$$[p]\,d\Omega = \frac{f^2 R^2 e^{-2g(t)}g'(t)dt}{\sqrt{E + f^2 R^2 e^{-2g(t)}}}. \tag{40'}$$

This is indeed exactly the decrease of the energy (38) during dt. Since all these connections hold universally for any kind of matter and, as I mentioned above without proving it, also for light or heat radiation, it must hold also for all that matter which supports the line element (28). Without having computed it, we may assert that the whole universe loses energy as if its contents worked by its pressure for increasing its volume.

What is this mysterious 'piston,' and where does the energy go to? In Newtonian mechanics one would say that it is spent to overcome the mutual gravitational attraction and stored as potential energy of gravity. From Einstein's theory the notions of gravitational pull and potential energy have disappeared, though they are used occasionally for brevity of speech. But are there not very important conservation laws at the basis of this theory, including the conservation of energy? Are they not violated if energy is said to diminish without there being a flow outwards (which there cannot be, because there is no boundary and no outside)? Well, no. The conservation law does not, allow one to assert that the energy content of any given spatial region is constant provided there is no energy traffic through the boundary; and that for the simple reason that the energy density is a tensor component (not an invariant) and its integral over an invariantly fixed region of space has no invariant or covariant meaning at all, not even that of a tensor – or, vector – *component*. It is true that in dealing with such a specialized object as an expanding spherical space one must greatly reduce one's demands on invariance, because one is tied to a very special frame; in a general frame the object would change its simple aspect and get completely out of hand. Non-invariant notions, adapted to the frame, may be very useful in such cases, a fact, by the way, that, is not always sufficiently recognized. Now in *any* special frame the conservation laws can be given a form which does allow integration over three-dimensional regions and assertions about 'constant content' of the kind mentioned above; the well-known trick is to turn the conservation laws into plain divergences with no accessory terms originating from the curvature. However, this is achieved at the cost of a non-invariant (pseudo-) tensor of energy-momentum-stress making its appearance along with the matter tensor proper. Only for the sum of the two do the integrated conservation laws hold. The thing that, in our case is independent, of time is the integral, over the whole space, of the *sum* of the material energy density and the pseudo-energy density of gravitation. And, of course, the latter contributes exactly what in Newtonian mechanics would be called the potential energy of gravitation. This is the solution of the apparent paradox, whether we like it or not.

From the preceding analysis one must not jump to the conclusion that the *rest* mass of the universe, because it is an invariant, is a constant. It is not, it can change; however, this is not directly connected with the change of volume, the expansion. We know that the rest mass changes in many processes in the laboratory, as in pair creation and so-called annihilation, and in all nuclear transformations. In these processes, which certainly are happening all the time on a large scale in the interior of the stars, *energy* is locally conserved but rest energy may change considerably. In these transitions, whenever the rest energy decreases, the kinetic energy increases and therefore the contribution of this particular amount of energy to the pressure increases, reaching its maximum value when the transition is to electromagnetic radiation with rest energy zero. R. C. Tolman has very aptly remarked that these changes of pressure when occurring in an initially static universe would by themselves suffice to upset the gravitational equilibrium and let expansion (or contraction) set in. Since we know that these processes are going on all the time, this is a very strong argument (in addition to, and independent of, the observed red shift) for believing that our universe is not static. For according to the accepted theory a change in the total rest energy must be accompanied by a change in the volume. There is not a unique relation between the two, but with constant volume the rest energy would have to be constant too.

(c) The red shift for spherical spaces

We shall inquire directly into the frequency shift of light, using the same method as before, i.e. equation (16)

$$\frac{\delta s_2}{\delta s_1} = \frac{\left[g_{ik}\frac{dx_i}{d\tau}\frac{\delta x_k}{\delta s}\right]_1}{\left[g_{ik}\frac{dx_i}{d\tau}\frac{\delta x_k}{\delta s}\right]_2} \tag{16}$$

the index 1 refers to the source, 2 to the observer, the δ-quotients to their respective world lines, the d-quotients to that of the light-ray. We take both source and observer to be without peculiar motion. We need the appropriate $d\tau$, which is given by our first integrals, or simply from (33) with $E = 0$:

$$d\tau = e^{g(t)}dt$$

(f may be dropped, it, cannot be zero). Only the term $i = k = 4$ survives; $g_{44} = R^2$ is a constant and can be dropped. The δ-quotients are also constants. Thus

$$\frac{\delta s_2}{\delta s_1} = \frac{e^{-g(t_1)}}{e^{-g(t_2)}}e^{g(t_2)-g(t_1)} = \frac{R(t_2)}{R(t_1)}. \tag{43}$$

The result is familiar to us from a previous case (the flat expanding Lemaître-de Sitter universe). The wave-lengths are now in exactly the relation of the radii of space at absorption and emission, respectively. Moreover, there is also full agreement with the slowing down of the momentum of a particle, if this momentum is interpreted 'for the photon' wave-mechanically as h/λ. The red shift is something that happens to the light on its journey, along with the expansion of space, not really a Doppler effect. But you may remember that in the Lemaître case we could alternatively interpret it as a Doppler effect on emission. This is obviously not possible in general (only perhaps when $g(t)$ is a linear function; but there are difficulties of thought in this interpretation anyhow; for the linear velocity of light is a constant and the velocity of recession of the source on emission (t_1) if you take the observer to be permanently at rest, is not the same as the velocity of recession of the observer (t_2), if you take the source to be permanently at rest. In any case, with a general $g(t)$ there is no general connection between the ratio of the two R-values at those two moments and their time derivatives at, one or the other or both these moments. Even in the case of the contracting and expanding spherical de Sitter world, where $R(t) \sim \cosh t$, there does not seem to be any simple connection with the 'velocities' of source and observer.

It is interesting to observe that the dependence of the change of wave-length on the change of volume of space is exactly the same as in a case well known from the classical deduction of Wien's displacement law: when you have any kind of light radiation enclosed in a perfectly reflecting enclosure and change the volume very slowly ('adiabatically'), e.g. by slowly drawing out (or pushing in) a perfectly reflecting piston, the wave-length changes proportionally to the third root of the volume. In that case one knows that the change is brought about by reiterated reflection at the receding or advancing piston. In the present case there is no piston! In the Wien case one also knows that the total energy must change by the amount of work done by the radiation pressure on the piston. Let V be the volume, u the average density and p the pressure, then since $p = \frac{1}{3}u$,

$$d(uV) = -pdV = -\frac{1}{3}udV,$$

$$\frac{d(uV)}{uV} + \frac{1}{3}\frac{dV}{V} = 0$$

uV constant. $V^{-\frac{1}{3}}$.

Thus the total energy changes in inverse proportion to the wave-length and therefore in direct proportion to the frequency. In our present case one cannot, from our simple considerations about geodesics, draw any justifiable inference about the energy change, whether it changes at all and, if so, how. Wave theory clears this point and actually confirms that the energy does vary in proportion to the frequency and therefore just by the amount of work done by radiation pressure against the imaginary piston. Wave theory demands that these changes of wave-length and energy shall occur for any isolated wave group traveling along anywhere in space. They are just a consequence of the Riemannian metric. Of course they can hold only inasmuch as our smoothed-out metric represents the actual universe. Moreover, they hold for an observer without peculiar motion in that smoothed-out universe, not for any other one, since wave-length and energy are not invariants.

III. Waves in General Riemannian Space-time

9. The nature of our approximation

What we have done up to here, using the geodesic theorem (or hypothesis), was purely geometrical optics and geometrical mechanics. It does not represent, our present views about the nature of light and of matter. We are therefore not so much interested in the question whether or no the geodesic assumption for test particles is a strict consequence of the field equations of gravitation, which it can at best be for material particles. We are more interested to know whether it is a consequence of the wave theories of light and matter. The answer is yes, even for a quite general gravitational field; but two points must be kept in mind. If the description as a wave phenomenon is correct, whether for light or for matter, the notion of particle orbits or light-rays is only an approximate conception, and that by intrinsic necessity (not perhaps by mathematical impotence), because the notions of frequency and wave-length (or wave-number), which are correlated to the energy and momentum of the particle, are only approximate notions–there are wave phenomena to whose structure they do not apply at all. So you must not expect anything but approximate theorems; to pretend them to be exact would be an imposture, and that on pure logical grounds. Secondly, the points of view of approximation are not generally invariant. They consist in assuming certain quantities to vary 'slowly' in space and time and others to be very nearly linear functions of the coordinates. Obviously such assumptions do not hold in an arbitrary frame. To make them amounts to admitting that there are frames in which they do hold. The range of these frames is still considerable. Invariance is not abandoned; but the prejudice must be abandoned, that a stipulation which does not hold in *every* frame of reference is meaningless. The bridge between particle theory and wave theory is here, as always, the Hamilton-Jacobi theory of particle motion. So we must, supply this first.

10. The Hamilton-Jacobi theory in a gravitational field

We have defined the 'paths' of free particles in a gravitational field by the variational principle

$$\delta I = \delta \int Q d\tau = 0, \qquad Q = g_{ik} \frac{dx_i}{d\tau} \frac{dx_k}{d\tau}, \qquad (1)$$

with the additional demand $Q = 0$ for null geodesics (lightrays). In the variation the limits of τ must be kept constant and the four functions $x_k(\tau)$ must not be varied at the limits. For time-like geodesics $d\tau$ may be identified with ds, making $Q = 1$. For light this is not so, yet $d\tau$ is determined on every null geodesic up to a constant multiplier. *We shall leave light aside at first and include it later.* We compute the variation of I keeping the four functions fixed at the initial point, but not at the end-point; one easily finds

$$\delta I = 2 g_{ik} \frac{dx_i}{d\tau} \delta x_k \bigg|_{\text{end-point}} . \qquad (2)$$

126

Since we have excluded the case of light ($Q = 0$) we may put $d\tau = ds$, the line element *on the non-varied, geodesic*; thus *on it* $Q = 1$ We wish to deduce from (2) the partial derivatives of the integral

$$I(x_k) = \int_{(x_k^0)}^{(x_k)} Q \, ds = \int_{(x_k^0)}^{(x_k)} ds, \qquad (3)$$

conducted from an arbitrarily fixed point (x_k^0) along a time-like geodesic to any point (x_k) inside the light-cone of (x_k^0), and regarded as a function of the x_k. We therefore use (2), taking for the varied path the time-like geodesic leading from (x_k^0) to $(x_k + \delta x_k)$ Still the change in $I(x_k)$ is not directly indicated by (2), because on the varied curve Q is not 1 but

$$Q = \left(\frac{ds'}{ds}\right)^2,$$

if ds' denotes everywhere the infinitesimal portion of the varied curve that is associated with the portion ds on the unvaried curve. We can obviously not make this ratio 1 every-where, but we can make it everywhere the same, viz. make

$$Q = \left(\frac{I(x_k + \delta x_k)}{I(x_k)}\right)^2$$

on the varied curve. Integrating this (with the parameter ds over the unvaried curve!) we obtain the meaning of δI in (2), thus

$$\delta I = \frac{[I(x_k + \delta x_k)]^2}{I(x_k)} - I(x_k) = 2\frac{\partial I}{\partial x_k}\delta x_k,$$

and therefore

$$\frac{\partial I}{\partial x_k} = g_{ik}\frac{dx_i}{ds}\bigg|_{\text{end-point}} \qquad (k = 1, 2, 3, 4) \qquad (4)$$

127

are the partial derivatives of the function (3). Since on any time-like geodesic $Q = 1$, it follows

$$g^{ik}\frac{\partial I}{\partial x_k}\frac{\partial I}{\partial x_i} = 1, \tag{5}$$

which is our Hamilton-Jacobi equation.

Conversely if we are given any solution of (5) and posit (4), the integral curves of (4) are geodesics. To prove this, differentiate (5) with respect to x_l use (4):

$$2g^{ik}\frac{\partial I}{\partial x_k}\frac{\partial^2 I}{\partial x_i\partial x_l} + g^{ik}_{,l}\frac{\partial I}{\partial x_k}\frac{\partial I}{\partial x_i} = 0,$$

$$\frac{dx_i}{ds}\frac{\partial}{\partial x_i}\left(g_{ik}\frac{dx_k}{ds}\right) + \frac{1}{2}g^{ik}_{,l}g_{lk}g_{ri}\frac{dx_l}{ds}\frac{dx_r}{ds} = 0, \tag{6}$$

$$\frac{d}{ds}\left(g_{ik}\frac{dx_k}{ds}\right) - \frac{1}{2}g_{ik,l}\frac{dx_i}{ds}\frac{dx_k}{ds} = 0;$$

this is the equation of the geodesic.

If one has a function $I(x_1, x_2, x_3, x_4, \alpha)$ which satisfies (5) identically in α then its derivative with respect to α is an integral of (4). Indeed, if (5) is differentiated with respect to α, then

$$0 = 2g^{ik}\frac{\partial I}{\partial x_k}\frac{\partial^2 I}{\partial x_i\partial\alpha} = 2\frac{dx_i}{ds}\frac{\partial^2 I}{\partial x_i\partial\alpha} = 2\frac{d}{ds}\left(\frac{\partial I}{\partial\alpha}\right). \tag{7}$$

If there are three such parameters $(\alpha_1, \alpha_2, \alpha_3)$, the integrals

$$\frac{\partial I}{\partial\alpha_k} = \beta_k \qquad (k = 1, 2, 3) \tag{8}$$

have sufficient (viz. 6) constants for fulfilling initial conditions. But naturally you can only get *time-like* geodesics. This can also be seen from (8). The direction of the normal to this hypersurface is, from (7), orthogonal to the direction of grad I, which according to (5) is time-like. Thus the

128

aforesaid normal is space-like. There are three such independent space-like normals to the curve of intersection of the three hypersurfaces (8), this intersection has therefore a space-like orthogonal R_3, it is thus time-like.

Notice that in these theorems there is no longer any reference to an initial point of integration, since I is now defined not by line integration but as a solution of (5). Moreover, the theorems that, we have just deduced for solutions I of equation (5) are self-contained, they do not depend on the way which has led up to this equation nor on the fact that the integral (3) satisfies it. Even if we knew nothing about this and had only just 'fancied' equation (5), we should find these two theorems true in exactly the way we did. It is not possible to *include* a null geodesic and it is not, possible to transfer to them the considerations that led up to (5), because the null geodesics that, start from a fixed x_k^0 define (3) (in which, of course ds would have to be replaced by the appropriate parameter $d\tau$) only on the light-cone of apex x_k^0, and define it as zero there. But it needs little imagination to 'fancy' the Hamilton-Jacobi equation for null geodesics, viz.

$$g^{ik}\frac{\partial I}{\partial x_k}\frac{\partial I}{\partial x_i} = 0. \tag{5'}$$

The two theorems for the solutions of (5) extend to those of (5'), except that the parameter τ must be used instead of s; thus in particular one has to put

$$g_{ik}\frac{dx_i}{d\tau} = \frac{\partial I}{\partial x_k} \tag{4'}$$

instead of (4), while (8) remains unchanged. One novel feature occurs, viz. if you differentiate I along an integral curve of (4'), then

$$\frac{dI}{d\tau} = \frac{\partial I}{\partial x_k}\frac{dx_k}{d\tau} = \frac{\partial I}{\partial x_k}g^{ik}\frac{\partial I}{\partial x_i} = 0. \tag{9}$$

129

Thus
$$I = \text{constant} \tag{10}$$
is now itself an integral of $(4')$. The path is in any case orthogonal to any member of the family (10) of hypersurfaces that, it meets. In the null case every path lies in one of them; each of these hyperrsurfaces then *consists* of null geodesics, in the same way as a light-cone does. But they are not, just light-cones, they are something much more general.

11. Procuring approximate solutions of the Hamilton-Jacobi equation from wave theory

We envisage the general wave equation
$$(\mathfrak{g}^{lk}\psi_{,k})_{,l} + \mu^2\psi\sqrt{-g} = 0. \tag{11}$$

The comma denotes ordinary differentiation, the Gothic is
$$\mathfrak{g}^{lk} = g^{lk}\sqrt{-g}.$$

The first term divided by $\sqrt{-g}$ is the generalized d'Alembertian of a scalar; the reason for including the second term, in which μ^2 is a constant, will appear presently. The wave function ψ is treated as a scalar. We choose it complex and split
$$\psi = Ae^{i\phi}(A, \phi \text{ real}) \tag{12}$$

To facilitate printing we shall drop the comma in the derivatives of A and ϕ. We have to insert (12) into (11):
$$\begin{aligned}\psi_{,k} &= A_k e^{i\phi} + iA\phi_k e^{i\phi}, \\ \mathfrak{g}^{lk}\psi_{,k} &= \mathfrak{g}^{lk}A_k e^{i\phi} + i\mathfrak{g}^{lk}A\phi_k e^{i\phi},\end{aligned} \tag{13}$$

$$(\mathfrak{g}^{lk}\psi_{,k})_{,l} = [(\mathfrak{g}^{lk}A_k)_{,l} + i\mathfrak{g}^{lk}A_k\phi_l + i(\mathfrak{g}^{lk}A\phi_k)_{,l} - \mathfrak{g}^{lk}A\phi_k\phi_l]e^{i\phi}.$$

We insert this and (12) in (11), drop the common exponential factor, then split into the real and imaginary parts:

$$(\mathfrak{g}^{lk}A_k)_{,l} - \mathfrak{g}^{lk}A\phi_k\phi_l + \mu^2 A\sqrt{-g} = 0. \qquad (14a)$$

$$\mathfrak{g}^{lk}A_k\phi_l + (\mathfrak{g}^{lk}A\phi_k)_{,l} = 0. \qquad (14b)$$

If the second equation, after a slight rearrangement, is multiplied by A, it reads

$$2\mathfrak{g}^{lk}AA_k\phi_l + A^2(\mathfrak{g}^{lk}\phi_k)_{,l} = 0,$$

thus

$$(A^2\mathfrak{g}^{lk}\phi_k) = 0; \qquad (15)$$

this will prove to be an extremely valuable *conservation law*. It is exact, no approximation has yet, been introduced.

But our procedure is only intended to give information for gravitational fields g_{ik} and solutions ψ for which the following hold:

(i) the relative changes of both the g_{ik} and the real amplitude A in any four-dimensional direction are very small compared to the changes of the phase ϕ that occur in some directions;

(ii) the phase ϕ is approximately a linear function of the four coordinates, in other words its second derivatives are small compared to the squares of its first derivatives.

According to (i) we divide (14a) by $A\sqrt{-g}$ and drop the first term (the ratio of the d'Alembertian of A and A itself). Then we have

$$g^{lk}\phi_k\phi_l = \mu^2. \qquad (16)$$

Thus the phase is (essentially) an approximate solution of the Hamilton-Jacobi equation (5) or (5'), according as

μ differs from zero or not. In other words, the orthogonal trajectories of the hypersurfaces of constant phase are (approximately) a family of geodesics.

Four-dimensional pseudo-orthogonality is not, easily visualized, especially in the null case. We shall therefore simplify the frame as much as possible at some point (P) of the wave and study the meaning of (15) and (16), and of their consequences according to the Hamilton-Jacobi theory in the neighbourhood of this point. At that point we make the metrical tensor strictly Galilean $(-1, -1, -1, 1)$, and stationary, i.e. all its derivatives zero. This will hold approximately in its neighbourhood, if we assume (i) and (ii) to hold in this frame as we shall. In what, follows I shall not repeat the restriction 'approximately' *ad nauseam* but'take it to be understood.

Through P passes a phase hypersurface and a geodesic. For the latter, according to the general theory

$$\frac{dx_k}{d\tau} = \eta^{kl}\phi_l \qquad (17)$$

(η is the Galilean tensor, ∓ 1; we use $d\tau$, not ds, in order to include the case of light, $\mu = 0$, and to swallow up the factor μ anyhow). Let, the phase hypersurface passing through P be

$$\phi(x_1, x_2, x_3, x_4) = C(= \phi(x_1^P, x_2^P, x_3^P, x_4^P)).$$

Then the phase surface passing at that moment, through P is,

$$\phi(x_1, x_2, x_3, x_4^P) = C. \qquad (18)$$

The first three equations (17) (for $k = 1, 2, 3$) tell us that the spatial orbit is normal to (18) in the plain Euclidean sense, and this holds in any case, i.e. also for light.

The proper meaning of the partial derivatives ϕ_l, is brought to the fore when we implement our assumption (ii), viz.

$$\phi = k_1 x_1 + k_2 x_2 + k_3 x_3 + k_k x_4 + k_5, \tag{19}$$

where the k_l are not, just constants, but slowly varying. Then

$$\phi_l = k_l + x_m \frac{\partial k_m}{\partial x_l} + \frac{\partial k_5}{\partial x_l}.$$

In the neighbourhood of P we only keep the first term and have for

$$l = \begin{matrix} 1 & 2 & 3 & 4 \end{matrix}$$
$$\frac{1}{2\pi}\phi_l = \begin{matrix} \alpha/\lambda & \beta/\lambda & \gamma/\lambda & \nu, \end{matrix} \tag{20}$$

with λ the wave-length, ν the frequency, α, β, γ, the direction cosines of the normal. The sign is taken care of, though one might have to reverse all four signs. Then from (16)

$$4\pi^2 \left(-\frac{1}{\lambda^2} + \nu^2 \right) = \mu^2,$$

and the phase velocity

$$c_{ph} = \lambda\nu = \sqrt{(1 + \mu^2\lambda^2/4\pi^2)}. \tag{21}$$

It is 1 for tight ($\mu = 0$). Otherwise there is dispersion. (The equation (21) is just de Broglie's dispersion formula if $\mu/2\pi$ is identified with the reciprocal of the Compton wave-length, which is $h/m_0 c$ in customary units.) The phase velocity is *not* the velocity of the particle on the geodesic (17). That is

$$\frac{\sqrt{(dx_1^2 + dx_2^2 + dx_3^2)}}{|dx_4|} = \frac{\sqrt{(d\phi_1^2 + d\phi_2^2 + d\phi_3^2)}}{|d\phi_4|} = \frac{1}{\lambda\nu},$$

thus simply the reciprocal of c_{ph}. It is noteworthy that it can be obtained alternatively by computing the *group velocity* in the ordinary way, viz.

$$c_{gr} = \frac{\partial \nu}{\partial(1/\lambda)} = \frac{\partial}{\partial(1/\lambda)} \sqrt{\frac{1}{\lambda^2} + \frac{\mu^2}{4\pi^2}} = \left(1 + \frac{\mu^2 \lambda^2}{4\pi^2}\right)^{-1/2}.$$

(22)

We turn to investigate the conservation law (15) in our simplified frame, thus *always in the neighbourhood, of our world point P*. It reads

$$(A^2 \eta^{lk} \phi_k)_{,l} = 0$$

(23)

or from(17)

$$\left(A^2 \frac{dx_l}{d\tau}\right)_{,l} = 0.$$

(24)

We write it *in extenso* thus:

$$\left(A^2 \frac{dx_4}{d\tau}\right)_{,4} + \left(A^2 \frac{dx_4}{d\tau}\frac{dx_1}{dx_4}\right)_{,1}$$

$$+ \left(A^2 \frac{dx_4}{d\tau}\frac{dx_2}{dx_4}\right)_{,2} + \left(A^2 \frac{dx_4}{d\tau}\frac{dx_3}{dx_4}\right)_{,3} = 0.$$

(25)

The τ-derivatives must be understood to be field functions, determined at any world point by our geodesic passing through that point. We are faced with the ordinary equation of continuity of a fluid (in *three* dimensions) of density $A^2(dx_4/d\tau)$ and streaming with the *particle* velocity. This interpretation lends itself particularly in the case $\mu \neq 0$, where we may take $d\tau = ds$, without loss of generality. There A^2 obviously plays the part, of *rest density*. With tight there is the inconvenience that there seems to be no *general* method of normalizing $d\tau$ *all over the place*, i.e. from one geodesic to the next. But in the form (23) the relation remains irreproachable: the four quantities inside the bracket are the components of a four-current, for which the equation of continuity holds.

134

It can be used in any case to determine the motion of a wave-parcel of the ordinary description, large compared to the wave-length, but still small enough to speak of its 'path' in the neighbourhood of the point P, the wave function being zero or inappreciable outside. The following integrals are three-dimensional space integrals over a region containing the parcel, but with $A = 0$ on the boundary. First we integrate (23) over this space, the parts for $l = 1, 2, 3$ vanish by partial integration, thus

$$\frac{d}{dx_4} \int A^2 \phi_4 dv = 0. \tag{26}$$

Next we multiply (23) by x_m $(m = 1, 2, 3)$ and write it

$$(A^2 x_m \eta^{lk} \phi_k)_{,l} - A^2 \eta^{mm} \phi_m = 0.$$

(In the second term $\eta^{mm} = -1$, and there is *no* summation.) On integration as before the terms $l = 1, 2, 3$ vanish, so

$$\frac{d}{dx_4} \int A^2 \phi_4 x_m dv = -\int A^2 \phi_4 \frac{\phi_m}{\phi_4} dv. \tag{27}$$

From (26) the $A^2 \phi_4$ fulfills the requirements of a reasonable weight function, since its space integral is a constant. The average x_m coordinate of the parcel moves at the rate of the average of

$$-\frac{\phi_m}{\phi_4} = \frac{dx_m}{dx_4} \tag{28}$$

(see (17)), both averages being taken with the same weight function. (They can be normalized by dividing them by the integral in (26).) If the three quantities (28) are fairly constant in the main part of the parcel, the parcel may be said to move with the particle velocity (or group velocity) along the wave normals or spatial orbits.

Though all our statements refer to a very special frame, they can be tested at any moment at, any world point by introducing this special frame, provided, our assumptions hold there. (If not, not.) They have therefore a more general validity than might seem at first sight. Such notions as the particle velocity (see (28)) are, of course, eminently non-invariant. But the statement that an imaginary mass point, put into 'the middle' of the parcel and subjected to the laws of the geodesic, remains covered by the parcel and more or less in its 'middle' is either true or false, but if true, it is independent of the frame.

Perhaps the following ought to be emphasized. Our proof that this statement holds approximately in certain conditions, which we have specified, does not in any way imply that a wave parcel will 'keep together' for ever, since the specified conditions need not and will not persist. Indeed, it is known that any wave parcel tends to spread indefinitely when you follow its course into the distant future or into the remote past.

IV. Waves in an Expanding Universe

12. General considerations

We now wish to study from the wave point of view the same phenomena that in Chapter II we have investigated from the point of view of moving particles or of geodesics. Before we start the analytical work, I wish to point out the nature of the problem and the conceptual difficulties that we are up against, lest you find the methods we have to adopt, too sophisticated. As we saw in Chapter III, it, is not, very difficult to set up, even in a quite general Riemannian space-time, the generalizations of d'Alembert's wave equation or of the one called the Klein-Gordon equation; for a more detailed study, the generalized field equations of Maxwell, Dirac, Proca could be dealt with similarly. In the relatively simple case of our expanding hypersphere it is not very difficult to obtain a complete survey of the solutions in all these cases. It is merely a matter of routine work, not very interesting in itself, very similar to the many other cases of solving such equations by separation of variables and spotting the several required sets of complete orthogonal functions of one variable, with respect, to which sets a series development of the general solution is obtained. The task is rendered a little involved, but also more interesting, by the appearance, in the equations, of *one* unspecified function $R(t)$, telling the radius of the hypersphere as a function of time; one tries, of course, to leave it unspecified as long as possible, because one wishes to obtain general results. However, all this is routine, quite involved routine at times, but the main point, and what really interests us is the changing configuration of a homogeneous wave group or wave parcel (whether of light or of matter waves) formed of nearly monochromatic waves, i.e. of approximately specified wave-length, and proceeding in (nearly) one direction. Now in most, if not, all, similar cases we are interested only in the velocity with which, and in the curve along which, the wave group proceeds, and we are satisfied to compute the former as the group velocity from the way in which the reciprocal phase velocity (or 'index of refraction') varies with the wave-length at the 'point' of the wave parcel, and to compute its path from the way in which this 'index of refraction" varies in the immediate neighbourhood of that 'point'. In the present case further considerable interest lies in the question how the whole configuration of the parcel changes on account of the radius of space $R(t)$ changing with time. This may and actually does produce a change of wave-length, which would otherwise be a constant, hyperrspherical space being homogeneous. The change of wave-length in turn entails a change of the velocity of propagation whenever there is dispersion, as there is at any rate for matter waves. This makes it necessary to pursue the wave parcel more carefully than is usually done.

In homogeneous flat space a typical wave parcel is obtained by superposing all the *infinite* plane waves whose wave normals lie within a slim cone of very narrow opening and whose wave-length is within a narrow interval) $\lambda, \lambda + \delta\lambda$. It turns out that by this superposition you actually obtain a small wave group, since everywhere outside a certain small region of space the constituent plane waves very nearly cancel each other by interference. In Riemannian space this synthetic construction is not, rigorously possible, *because there are no plane waves filling the whole of space.* We meet with the same difficulty in the optics of inhomogeneous media as well as in the wave mechanics of conservative systems, which is of even more general interest in physics. Here configuration space is also Riemannian and quite often genuinely Riemannian (e.g. in the case of the motion of a rigid body), not merely for our having chosen curvilinear coordinates. In all these cases the difficulty is quite duly waived by considering the above construction (composition of plane waves) to be made only in the neighbourhood of the point in question and by disregarding the rest of space or configuration space. But our present case is different because the Riemannian structure of space varies with time, though in a comparatively simple way. The wave-mechanical treatment of non-conservative systems has to face exactly the same situation, but it has never reached great prominence.

There is not anything like a plane wave filling the whole hypersphere homogeneously. At first sight this is astonishing, for there are *movements*, very much like 'translations', of the hypersphere in itself: all points moving at the same rate, each alone a great circle. (This is not so for the common sphere of two dimensions; it depends on the parity of the number of dimensions; for the one-dimensional sphere, the circle, it is obviously true.) One wonders, whether there could not be on the hypersphere a progressing wave with wave fronts orthogonal to the great circles that are the paths of such a 'translatory' movement? They do fill the whole hypersphere; but they are twisted or interlocked: if you contracted (on the hypersphere) one of them into a point, it would during this performance cut, through *all* the others. They cannot form the rays of a wave motion because they have not what, is often called the 'ray property', i.e. they have no surfaces orthogonal to them.

To overcome our difficulty *two ways* seem to offer. One is to construct a wave parcel that obeys the wave equation (modified for curvature and expansion) *with sufficient accuracy* in the spatio-temporal vicinity of one world point, and to see how it behaves during a short interval of time. We have dealt with this in the last chapter, but it is not very satisfactory *for two reasons.* Within a small region the influence both of curvature and of expansion is extremely slight. Thus even an *unmodified* plane wave of constant frequency obeys the modified wave equation very very closely, yet, of course, not with sufficient accuracy; for it just fails to exhibit the phenomena we are out for. We should have to improve it. But with an excellent approximation in hand it is not easy to know how far one has to improve it for obtaining the fullest possible information. And here we encounter the second reason for dissatisfaction. We need information holding not only for a short time, but for long periods, such as the time that a light-signal takes to cover the distance between a faint extra-galactic nebula and ourselves, which is probably a substantial fraction of the circumference of space. On the basis of an approximate solution such information in the large is only obtainable by very careful scrutiny, and even so not in a simple, easily surveyable form.

We shall therefore now choose the *second method,* though at first sight it seems to aim at much more than we really need. It consists in determining in principle all the proper vibrations of our expanding universe. From them certainly any wave group can be composed. One is inclined to consider this a sophisticated detour. But observe that it is the straight-forward generalization of what one always does in static flat space, whose proper modes are precisely the plane sinusoidal waves. The proper modes of a pseudo-sphere are less simple. In particular, space being finite and closed, one cannot by direct inspection of the spatial form of every eigen-function accord it, a *wave-length* – as is well known from the vibrations of finite bodies, rectangular or spherical enclosures, etc. But this is of no relevance.

13. Proper vibrations and wave parcels

As in Chapter III, § 11, we want to investigate the phenomena which in flat space-time would be controlled by the second-order equation

$$-\nabla^2 \psi + \frac{1}{c^2} \frac{\partial^2 \psi}{\partial t^2} + \mu^2 \psi = 0, \qquad (1)$$

for the scalar wave function ψ. The constant c is the phase velocity of light in flat space-time. For $\mu = 0$ this is just d'Alembert's equation, while for

$$\mu = \frac{2\pi m_0 c}{h} = \frac{2\pi}{\lambda_c} \qquad (2)$$

(λ_c being the Compton wave-length for a particle of rest mass m_0), equation (1) is called the Klein-Gordon equation for a free particle of mass m_0. It would in many respects be preferable to use in both cases, instead of (1), the set of linear equations from which (1) is the eliminant, viz. Maxwell's equations for light, Dirac's for particles with spin $\frac{1}{2}$, etc. Even the mathematical treatment is in some cases simpler – but only in principle. In actual fact a wave function with many components is much more difficult, to handle and to visualize. Moreover, the set of first-order equations is different for particles with different spin numbers, so the task immediately splits up into at least two or three mathematically different problems, and those referring to half-odd-integral spin are rather intricate. For the handling of spinors, properly speaking, in a Riemannian space-time is not a quite straight-forward affair. One

141

has first to decide on their general transformation laws, which are not given in ordinary tensor analysis. I do not think there is any real ambiguity in this decision; but it is much less general than ordinary tensor analysis, because the very notion of spinors is so intimately connected with the four-dimensional Lorentz group. That is why for a first orientation equation (1) is to be preferred.

There is no doubt about, its generalization which was already used in §11. The μ^2 term is by itself an invariant, the other two terms are the (four-dimensional) divergence of the gradient in the special metric characterized by the line element.

$$ds^2 = g_{ik}dx_i dx_k = -dx_1^2 - dx_2^2 - dx_3^2 + c^2 dt^2 \quad (3)$$

(we use the letter t for x_4). In a general relativity metric this reads

$$(-g)^{-1/2} \frac{\partial}{\partial x_k} \left[\mathfrak{g}^{ik} \frac{\partial \psi}{\partial x_i} \right] + \mu^2 \psi = 0. \quad (4)$$

We wish to specialize in a general expanding universe:

$$ds^2 = -R(t)^2 d\eta^2 + c^2 dt^2. \quad (5)$$

Here $d\eta$ is short for the line element on the three-dimensional unity hypersphere which we previously used in the form

$$d\eta^2 = d\chi^2 + \sin^2 \chi \, (d\theta^2 + \sin^2 \theta d\phi^2). \quad (6)$$

In (5) we have adopted two slight changes in notation, first, by writing $R(t)$ for what in previous work was called $Re^{g(t)}$; I beg to warn the reader that R, is from now on a function of t, even if the argument is not written explicitly. To avoid confusion and also for analogy with (1) and (3), we have given dt^2 the factor c^2 (not the constant R^2 as before). For actual work a different, form of (6) is a little more convenient, viz.

$$dη^2 = dω^2 + \sin^2 ω dα^2 + \cos^2 ω dβ^2. \tag{7}$$

It results from expressing the coordinates on the three-dimensional unity hypersphere

$$y_1^2 + y_2^2 + y_3^2 + y_4^2 = 1, \tag{8}$$

as follows

$$
\begin{aligned}
y_1 &= \sin ω \cos α, & y_3 &= \cos ω \cos β, \\
y_2 &= \sin ω \sin α, & y_4 &= \cos ω \sin β.
\end{aligned}
\tag{9}
$$

The pleasing symmetry of this allotment makes (7) a little simpler than (6), inasmuch as the coefficients depend only on *one* angle, viz. on $ω$. The angles $ω, α, β$, are duly termed cylindrical coordinates. The first ranges from 0 to $\frac{1}{2}π$, the other two from 0 to $2π$. The surfaces $ω$=constant are 'cylinders' (also called Clifford surfaces, with the connectivity of tori!); with Euclidean cylinders they share the property of having, each of them, a strictly Euclidean metric; their 'generatrices' are the two families of great circles $α + β=$ constant and $α - β =$ constant.

In order to form (4), we write out the elements of the metrical tensor, from (5) and (7)

$$
\begin{aligned}
g_{ik} &= (-R^2, \ -R^2 \sin^2 ω, \ -R^2 \cos^2 ω, \ c^2), \\
\sqrt{-g} &= cR^3 \sin ω \cos ω, \\
g^{ik} &= (-1/R^2, \ -1/(R^2 \sin^2 ω), \ -1/(R^2 \cos^2 ω), \ c^{-2}).
\end{aligned}
\tag{10}
$$

We thus obtain from (4)

$$- R^{-2} K(ψ) + c^{-2} R^{-3} \frac{∂}{∂t}\left(R^3 \frac{dψ}{dt}\right) + μ^2 ψ = 0, \tag{11}$$

143

where

$$K(\psi) = \frac{1}{\sin\omega\cos\omega}\frac{\partial}{\partial\omega}\left(\sin\omega\cos\omega\frac{\partial\psi}{\partial\omega}\right)$$
$$+ \frac{1}{\sin^2\omega}\frac{\partial^2\psi}{\partial\alpha^2} + \frac{1}{\cos^2\omega}\frac{\partial^2\psi}{\partial\beta^2}. \tag{12}$$

This lends itself to the familiar method of separation of variables. We put

$$\psi = Y(\omega, \alpha, \beta)f(t), \tag{13}$$

and demand that (12), formed of Y, be a constant, multiple of Y. This constant is $-l(l+2)$, where $l = 0, 1, 2, 3, \ldots$ for these are the eigen-values of the operator K. It is the operator of 'three-dimensional spherical harmonic', studied already by Legendre and in other classics on extending the notion of harmonic function (or potential theory) to more than three dimensions. In our variables the dependence on α and β is particularly simple, because it has to be periodic with period 2π. Thus

$$Y(\omega, \alpha, \beta) = S(\omega)e^{i(n\alpha+m\beta)}, \tag{14}$$

where m, n, are integers, and S is subject to the ordinary differential equation of the second order

$$\frac{1}{\sin\omega\cos\omega}\frac{d}{d\omega}\left(\sin\omega\cos\omega\frac{dS}{d\omega}\right) - \frac{n^2}{\sin^2\omega}S - \frac{m^2}{\cos^2\omega}S = kS. \tag{15}$$

(We do not use the knowledge about k quoted above.) This equation is singular at both border points $\omega = 0, \frac{1}{2}\pi$. At such a point *one* of the two independent solutions remains regular, but not usually the same at the two points. We must, of course, demand regularity everywhere and that is what imposes a condition on the constant k. Some of these solutions are very simple, viz.

$$S = \sin^{|n|}\omega\cos^{|m|}\omega, \tag{16}$$

144

belonging to the eigen-value

$$
\begin{aligned}
k &= -(|m| + |n|)(|m| + |n| + 2) \\
&= -l(l+2),
\end{aligned} \tag{17}
$$

say. From them others can be obtained which, given m and n, belong to any integer $l > |m| + |n|$. We shall not bother to determine them now (nor to prove that the eigen-values all have this form: product of two consecutive even or odd integers).

But let us try to visualize the solutions

$$
e^{i(n\alpha + m\beta)} \sin^n \omega \cos^m \omega \qquad (n \geq 0, m \geq 0)
$$

(to be supplemented by a function of the time which is very nearly $e^{2\pi i \nu t}$ and does not interest us at the moment). They are quite interesting. It is obvious that m and n (or at, least one of them) must be extremely large for obtaining a wave-length of 'human' dimensions. (We shall determine the wave-length exactly.) The function of ω is the amplitude. It shows a queer behaviour. Let

$$
A = \sin^n \omega \cos^m \omega \text{ and } \log A = y :
$$
$$
y = n \log \sin \omega + m \log \cos \omega,
$$
$$
y' = n \cot \omega - m \tan \omega.
$$

Since ω ranges only from O to $1/2\pi$, A has just *one* extremum, which must be a maximum, since A vanishes at the end-points. Call that value ω_0,

$$
\tan \omega_0 = \sqrt{n/m}.
$$

For large n and/or m this maximum is extremely sharp. Indeed

$$
y'' = -\frac{n}{\sin^2 \omega} - \frac{m}{\cos^2 \omega} = -n(1 + \cot^2 \omega) - m(1 + \tan^2 \omega),
$$

145

and at the place of the maximum ($\omega = \omega_0$)

$$y'' = -n\left(1+\frac{m}{n}\right) - m\left(1+\frac{n}{m}\right) = -2(m+n)\ (n \neq 0, m \neq 0),$$

and therefore with good approximation

$$A = A_0 e^{-(m+n)(\omega-\omega_0)^2}.$$

The whole wave event is confined to one of the torus cylinders, the angular width being $1/\sqrt{m+n}$ the linear width $R/\sqrt{m+n}$. We shall inspect this further, but first we wish to analyse the waves 'on this surface'. Remember

$$ds^2 = -R^2(d\omega^2 + \sin^2\omega\,d\alpha^2 + \cos^2\omega\,d\beta^2) + c^2 dt^2.$$

We are interested at the moment in the space part only, and in the torus $\omega = \omega_0$ (and its immediate neighborhood). On every such torus the metric is exactly Euclidean. We can cut, it open by two cuts along a 'meridian' and a 'parallel', and spread it on the plane of the paper; but the 'meridians' and 'parallels' play perfectly equivalent roles, they are the lines $\alpha = $ constant, $\beta = $ constant, respectively. Since on one of the first β goes from 0 to 2π, all the first have the length $2\pi R \cos\omega_0$, the second $2\pi R \sin\omega_0$. (Our drawing assumes that $\sin\omega_0 < \cos\omega_0$ thus $n < m$.) Notice first that opposite points on opposite edges are equivalent, i.e. mean the *same* point; more generally speaking, two points are equivalent if and only if their α's differ by integral multiples of 2π *and* the same holds for their β's. Moreover, notice the following. Though all straight lines are, of course, geodesics on the *torus*, they are not all geodesics on the *hypersphere* (as is immediately seen for the parallels and meridians, which are *not* great circles, since they have total length $2\pi R \sin\omega_0$ and $2\pi R \cos\omega_0$ respectively). But there are two families of great circles on our surface, namely, the two sets of lines, each set parallel

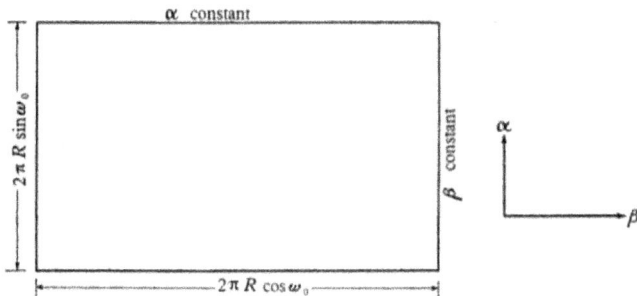

Fig. 9 The torus $\omega = \omega_0$ cut open and spread in the plane.

to a diagonal of the rectangle, respectively, for they have the length $2\pi R$.[1] Analytically they are characterized by α and β changing at the same rate

$$\frac{d\alpha}{d\beta} = \pm 1.$$

Now let us investigate the lines of constant phase on the surface, viz.

$$m\alpha + m\beta = k \quad \text{(a constant)}.$$

Do they close? Yes, but, in general only after many turns. For to get, back to the same point, both α and β must, increase by integral multiples of 2π, say $2\pi s$ and $2\pi r$, such that $ns + mr = 0$. If m and n have a common divisor, take the greatest out, and you remain say with $n's + m'r = 0$. Then $s = m', r = -n'$ are the smallest numbers of turns for returning to the same point.

The normals to the phase lines, the rays, are not found by forming the gradient with respect to α and β, because these have not the same values in length. But let $\delta\alpha$, $\delta\beta$ be the increments along the phase line:

$$n\delta\alpha + m\delta\beta = 0, \qquad \frac{\delta\beta}{\delta\alpha} = -\frac{n}{m} = -\tan^2 \omega_0.$$

---

[1] Also an ordinary torus carries two families of *congruent* circles, less familiar than its meridians and parallels.

147

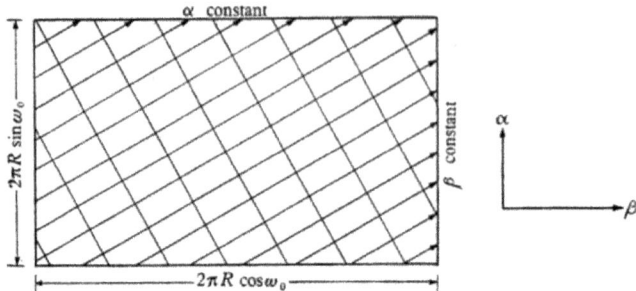

Fig. 10 Rays and wave surfaces on the torus; the rays distinguished by arrows are parallel to the diagonal of the rectangle.

For the element $d\alpha, d\beta$, to be orthogonal to $\delta\alpha, \delta\beta$, we gather from the line element the condition

$$\sin^2 \omega_0 d\alpha\delta\alpha + \cos^2 \omega_0 d\beta\delta\beta = 0,$$

$$\frac{d\alpha}{d\beta} = -\frac{\delta\beta}{\delta\alpha} \cot^2 \omega_0 = 1.$$

Hence the rays are one of our families of great circles, which is not very astonishing. Along any ray, whose length is $2\pi R$, α increases by 2π, and so does β. Hence the phase increases by $m + n$ times 2π, thus the wave-length λ is

$$\lambda = \frac{2\pi R}{m + n}.$$

We have just seen above that in general a phase-line spirals many times around the torus before it closes. How often does it cut a particular ray, e.g. $\alpha = \beta$? We must ask how often the difference $\alpha - \beta$ becomes a multiple of 2π along the whole closed wave line. Now we have seen that α *increases* by $2\pi m'$ and β *decreases* by $2\pi n'$ where m', n' are the m and n, freed from any possible common divisor. The difference $\alpha - \beta$ therefore increases by $2\pi(m' + n')$, and will pass through a multiple of 2π just $m' + n'$ times. This is the number of points of intersection.

148

We have the rather remarkable result that if m, and n have no common divisor, there is only one wave line for every particular phase, i.e. all places with the same phase lie on the same wave line, which manages to spiral around the whole torus *many times* (both in α and in β!) without cutting itself. If, however m and n have the greatest common divisor d there must be d wave lines for every phase. If $m = n$ (i.e. $\omega_0 = \frac{1}{4}\pi$), the rectangle becomes a square and the wave lines are the great circles parallel to the *other* diagonal (in our plane figure). Still every wave line cuts every ray *twice* $(m' + n' = 2)$ since they are *circles*. But I must give the warning that the general statements make good sense only if *both* m and n are fairly big, so that there is a practically field-free space both 'inside' and 'outside' the torus $\omega = \omega_0$.[2]

For the linear spread of the wave phenomenon orthogonal to the torus (in the ω-direction) we found $R/\sqrt{(m+n)}$. Now that we know the wave-length λ we can express it thus

$$\frac{R}{\sqrt{(m+n)}} = \sqrt{\frac{\lambda R}{2\pi}}.$$

So the thickness of our skin wave is, in order of magnitude, the geometrical mean between the wave length and the radius of space. The dependence on time, $f(t)$, is in all cases controlled by (11) with (13):

$$R^{-2}l(l+2)f + c^{-2}R^{-3}\frac{d}{dt}\left(R^3\frac{df}{dt}\right) + \mu^2 f = 0. \qquad (18)$$

We introduce a new time variable τ by

$$d\tau = R^{-3}dt, \qquad (19)$$

---

[2] The inverted commas mean that tho allotment of these terms is arbitrary.

149

and obtain

$$\frac{d^2 f}{d\tau^2} = -[c^2 l(l+2)R^4 + c^2 \mu^2 R^6]f. \tag{20}$$

The two independent, solutions of this equation indicate the dependence on time for any proper vibration belonging to the integer l. We note that they are not, strictly harmonic functions as they are in most eigen-value problems, and that for two reasons. First our variable τ is in the long run not, strictly proportional to t; secondly, in (20) the factor of f still depends on τ. However, we must consider that the phenomena we wish to study consist in very rapid oscillations compared to which the radius $R(t)$ varies slowly, and this is true whether we measure both time rates in the variable t or in the variable τ. We may therefore look upon (20) as controlling a phenomenon of very nearly harmonic vibrations, a pendulum for instance, with very slowly varying constants, e.g. for the pendulum its length or the constant of gravity. We may say that the vibration is harmonic, but with a very slowly changing frequency,

$$\nu' = \frac{c}{2\pi} \sqrt{l(l+2)R^4 + \mu^2 R^6} \tag{21}$$

(with respect to the variable τ), or

$$\nu = \frac{c}{2\pi} \sqrt{\frac{l(l+2)}{R^2} + \mu^2} \tag{22}$$

(with respect to the variable t). This view is justified provided that the relative change of $R(t)$ is very small during a period ν^{-1} of the waves.

What is the wave-length? In the case of any of the more complicated proper modes $Y(\omega, \alpha, \beta)$ in (13) this cannot be decided by direct inspection of the function. However, we know that from the entirety of all these functions any possible wave phenomenon can be composed. We

150

are justified in admitting that in this way one can *inter alia* obtain a small wave parcel of very nearly constant wavelength, and, proceeding in a very nearly definite direction, a parcel small compared to R, but, of course, large compared to the wave-length. Notice at once that this mathematical construction is the same for any kind of wave, whether light ($\mu = 0$), or associated with a particle of given non-vanishing rest mass, characterized by a particular value of μ. For the angular functions Y do not depend on μ, which only enters the equations that govern the time dependence, (18) or (20).

We are further justified in assuming that, for building up such a wave parcel only a narrow range of frequencies will be required. From (22) this means that only the proper modes within a narrow range of the integer l are required. Hence the wave-length must be determined by l and vice versa, irrespective of the kind of wave (the μ value), provided we exclude the possibility that for a given μ the correspondence between wave-length and frequency may not be one-one.

To obtain the relation between the wave-length λ and the integer l we may draw on our having introduced c as the phase velocity of light. Then from (22), for $\mu = 0$, the wave-length λ must be

$$\lambda = 2\pi R/l. \tag{23}$$

(This is sufficiently accurate, since l is then a huge number! After what has been said before, this must hold for any μ, i.e. for any kind of wave.) But, of course, our claim for c to be the velocity of tight is really a borrowing from the theory of null geodesics.

A more direct way of obtaining the relation (23) is to inspect one of the special Y functions that we have determined, viz,

$$Y = \cos^m \omega\, e^{im\beta} \qquad (m > 0, l = m). \tag{24}$$

151

Since m is to be a huge number this function is very sharply restricted to the great circle $\omega = 0$, fading away exponentially as one recedes from it. Indeed

$$\cos^m \omega \approx \left(1 - \tfrac{1}{2}\omega^2\right)^m \approx e^{-\frac{1}{2}m\omega^2}. \qquad (25)$$

The angle measured along this great circle is β. As it proceeds from zero to π, the last factor in (24) goes through m periods, which means that there are rn (or l) wavelengths along $2\pi R$, in accordance with (23). We notice by the way that the angular width of the thread-like function (25) is of the order $m^{-1/2}$, its linear width $Rm^{-1/2}$; this is large compared to the wave-length, small compared to the radius R, just about their geometric mean. By joining the time factor to Y of (24) and superposing the proper modes of a moderate m-interval we can build up a wave parcel that is also longitudinally limited, though its lateral spread, as we just pointed out, is not, as small as it might be with regard to its wave-length.

The most relevant fact that emerges from our almost, trivially simple statement (23) is that all wave-lengths expand along with the radius of space at, the same rate. For particles it means the loss of momentum that we have found before by the method of geodesics (see §8 (b)), for light it means the famous red shift. It is seen to be a change that happens on the journey, all along the path; to regard it as a Doppler effect on emission is (a) egocentric; the fellow at the other end would say that it happens on reception, because we, the observers, are moving backwards; and (b) it is not really correct, what really matters is the ratio of, say, $R(t_1)$ and $R(t_2)$.

The secular variation of the wave-length refers primarily to the single proper mode and is a simple and clear-cut affair; since the wave function is a product of a function of the angles and a function of the time, the whole angular pattern remains the same for all time. For a moment one

152

might think that the same would hold for a small parcel of progressing waves, formed by superposition of proper modes. But we are not entitled to this inference, because the time function is not exactly the same for all the constituent proper vibrations; and besides the inference would be patently false, since the wave packet in addition to (possibly and very probably) expanding also progresses; nay, the propagation is the much more rapid and obvious change, it is not a secular variation; within a very short time other and other angular districts are occupied, those previously occupied coming to rest. On account of this rapid change, which is the main feature, it is not so simple to demonstrate for the wave packet the slow secular *expansion*, which, however, will turn out to be a correct guess.

We first make out the secular change of the time function $f(t)$ attached to a single proper mode. It is controlled by (20). I have pointed out before that this is the equation for the elongation of a pendulum, or any other one-dimensional harmonic oscillator, with very slowly varying constants. We can apply to it a very general law about 'adiabatic variation' due to Paul Ehrenfest. It states in our particularly simple case that the *energy* of the pendulum, i.e. the square of the product, of its frequency and its amplitude, undergoes a secular variation proportional to the secularly varying frequency. In the variable τ this is ν', given by (21). Hence we have

$$\nu'^2\overline{|f|^2} \sim \nu'. \tag{26}$$

Replacing ν' by ν, the actual frequency of the proper vibration, we get, comparing (22) and (21),

$$\overline{|f|^2} \sim 1/(\nu R^3). \tag{27}$$

This law of variation is the same for all proper modes that have the same frequency, and is therefore very nearly the same for all those that compose our small wave packet. All the functions f in question are therefore simple harmonic functions, say $|f|e^{\pm i\nu t}$, with very nearly the same frequency, both the frequency and the amplitude $|f|$ undergoing a very slow secular variation, that of the amplitude being given by (27). Hence it wilt suffice to take the time average just for one period ν^{-1}. Now let ψ be the wave function anywhere inside our wave packet. We think it developed into the series of proper modes, and we multiply by the complex conjugate series. All squares and products of amplitudes exhibit the secular variation indicated by (27). The same must hold for the whole product, thus

$$\overline{|\psi|^2} \sim 1/(\nu R^3), \qquad (28)$$

the bar referring, as mentioned above, to a time average over one period ν^{-1}. (This deduction is open to criticism; but since the result is correct, let it pass for the moment.) Now we compute the space integral of $\overline{|\psi|^2}$ in two ways. First without using (28), directly from the aforesaid series, which we integrate over the whole of space. On account of the mutual orthogonality of the angular functions, only the squares survive, and we obtain quite rigorously that the result is proportional to the product of the second member of (27) and the volume of space, thus

$$\int \overline{|\psi|^2} dV \sim \frac{1}{\nu R^3} R^3 = \frac{1}{\nu}. \qquad (29)$$

Secondly we use (28) and integrate only over the wave packet. The result is proportional to the product of the second member of (28) and the volume, say V, of the wave packet. Since both results must assert the same, we have

$$V/(\nu R^3) \sim 1/\nu, \qquad V \sim R^3. \qquad (30)$$

Hence obviously the dimensions of the wave packet increase along with the radius of space.

The result may be checked and made very obvious, if we grant, that in spite of the curvature of space a small 'optimal' wave parcel may be built up of plane waves, and allow for the secular change in wave-length. It is known that, the product of the longitudinal spread (say Δx) of the parcel and the spread of the wave number (say $\Delta(1/\lambda)$) is unity (or some pure number, according to definition). Now

$$\Delta\frac{1}{\lambda} = -\frac{\Delta\lambda}{\lambda^2} \sim R^{-1}, \tag{31}$$

hence Δx increases proportionally to R. Again let $\Delta\gamma$ be the constant angular spread of the wave normals. The lateral spread (say Δy) of the parcel is inversely proportional to $\Delta\gamma/\lambda$, and thus again directly proportional to R. The two relations that we have used, without proving them, form the well-known wave-mechanical foundation of the uncertainty principle, which results if you multiply them by Planck's constant h and call h/λ the momentum. The objection that plane waves are not available on the hypersphere may be met by using the angular proper modes (24), which are the best available substitute.

We wish to inquire now into the velocity with which the whole wave group proceeds, the so-called group velocity. As I have pointed out this is the much more rapid and prominent change that it undergoes. A rough orientation is easily obtained (in point of fact it turns out to be very accurate). We return to equation (22), giving the connection between the frequency ν and the number l which determines the wave-length, being the number of waves per circumference $2\pi R$ of space. Since this is *very* large we may in (22) read l^2 for $l(l+2)$, thus

$$\nu = c\sqrt{(1/\lambda_C^2 + 1/\lambda^2)}, \tag{32}$$

where λ_C, is the Compton wave-length (see equation (2)). This is on the one hand the de Broglie dispersion formula for 'matter waves', going over into that for light when $\lambda_C \to \infty$ (i.e. rest mass $m_0 \to 0$). On the other hand, if you multiply it by Planck's constant h and regard $h\nu$ as the energy, say W, of the particle and h/λ, as its momentum, say p, then (32) reads

$$W = \sqrt{(m_0^2 c^4 + p^2 c^2)}, \qquad (33)$$

which is the same as (38) (Chapter II), only in different notation. If from (32) we derive the group velocity in the familiar way

$$c_{gr} = \frac{d\nu}{d(1/\lambda)} = \frac{c/\lambda}{\sqrt{1/\lambda_C^2 + 1/\lambda^2}}, \qquad (34)$$

it, turns out to be the particle velocity v on account of the old relativistic connection, from (33),

$$\frac{dW}{dp} = \frac{c^2 p}{W} = \frac{c^2 m v}{m c^2} = v. \qquad (35)$$

(Here m is the mass, not the rest mass, which was m_0.) The phase velocity is obtained directly from (32)

$$c_{ph} = \nu\lambda = c\lambda\sqrt{1/\lambda_C^2 + 1/\lambda^2}. \qquad (36)$$

This with (34) gives the familiar relation

$$c_{gr} c_{ph} = c^2. \qquad (37)$$

The error committed on replacing $l(l+2)$ by l^2 is exceedingly small and could easily be estimated. But there is a graver objection against the above simple treatment, viz. that we have entirely disregarded the curvature of space and its expansion. Thus both λ and ν are not constants but slowly varying functions of the time; moreover, the orbit is an *expanding* great circle. Since for matter waves there is dispersion, the secular change of wave-length has the effect that both the phase velocity and the group velocity are subject to secular variation, corresponding exactly to the slowing down of particles that we found before by the method of the geodesics. This is, of course, rendered faithfully by our formulae (34) and (36) above, but a, more rigorous deduction might be desirable.

However, a completely rigorous one is lengthy, and since a half-rigorous one is not, better than the above, I shall not bother you with either here. Just to give you an idea of the kind of change one finds I shall write out the emendated dispersion formula (22) for a linear variation of the radius

$$R(t) = a + bt \tag{38}$$

with a and b constants. One obtains

$$\nu = \frac{c}{2\pi} \sqrt{\frac{l(l+2) - (b/c)^2}{R^2} + \mu^2}. \tag{39}$$

Even this is not entirely exact. Debye's asymptotic expansion for a Bessel function of extremely high order ($\sim cl/b$) and *comparably* great argument has been used. It will be seen that the influence is negligible, unless space expands with a *great multiple* of the velocity of light.

Bibliography

Eddington, A. S. (1930). *The Mathematical Theory of Relativity*, 2nd ed. Cambridge University Press.[1]

Einstein, A. (1950). *The Meaning of Relativity*, 3rd ed. Princeton University Press.[2]

Laue, M. von (1931). *Berliner Sitzungsberichte*, p. 123.

Laue, M. von (1953). *Die Relativitätstheorie*, II, 3rd ed. Braunschweig: F. Vieweg & Sohn.

Lemaître, G. (1925). *J. Math. Phys.* (M.I.T.), **4**, 188; *Phys. Rev.* **25**, 903.

Levi-Cività, T. (1927). *The Absolute Differential Calculus* (translated from the Italian). London and Glasgow: Blackie and Son.

Robertson, H. P. (1928). *Phil. Mag.* **5**, 835.

---

[1] EDITOR'S NOTE: New publication: A. S. Eddington, *The Mathematical Theory of Relativity*, With a Foreword by Abhay Ashtekar. Edited by Vesselin Petkov (Minkowski Institute Press, Montreal 2016).

[2] EDITOR'S NOTE: New publication in: A. Einstein, *Relativity: Meaning and Consequences for Modern Physics and for our Understanding of the World*. With an Introduction by Tian Yu Cao. Edited by Vesselin Petkov (Minkowski Institute Press, Montreal 2021)

Schrödinger, E. (1938). Eigenschwingungen des sphärischen Raums. *Comment. Vatican Acad.* **2**, 321.

Schrödinger, E. (1939). The proper vibrations of the expanding universe. *Physica,* **6**, 899.

Tolman, R. C. (1934). *Relativity, Thermodynamics and Cosmology.* Oxford University Press.

Weyl, H. (1922). *Space-Time-Matter* (translated from the German). London: Methuen.[3]

---

[3]EDITOR'S NOTE: New publication: H. Weyl, *Space-Time-Matter.* Edited by Vesselin Petkov (Minkowski Institute Press, Montreal 2021)